Fevered

Fevered

WHY A HOTTER PLANET WILL HURT OUR HEALTH—AND HOW WE CAN SAVE OURSELVES

LINDA MARSA

RODALE.

Mention of specific companies, organizations, or authorities in this book does not imply
endorsement by the author or publisher, nor does mention of specific companies, organizations,
or authorities imply that they endorse this book, its author, or the publisher.

Internet addresses and telephone numbers given in this book were accurate
at the time it went to press.

Rodale books may be purchased for business or promotional use or for special sales.
For information, please write to: Special Markets Department,
Rodale Inc., 733 Third Avenue, New York, NY 10017.

Printed in the United States of America

Rodale Inc. makes every effort to use acid-free ∞, recycled paper ♻.

Portions of this book were previously published in *Discover, Pacific Standard*, and in the
collection *Best American Science Writing, 2012*.

Book design by Amy C. King

Library of Congress Cataloging-in-Publication Data is on file with the publisher.

ISBN-13: 978–1–60529–201–4 hardcover

Distributed to the trade by Macmillan

2 4 6 8 10 9 7 5 3 1 hardcover

We inspire and enable people to improve their lives and the world around them.
rodalebooks.com

To William R. Roberts, who makes all things possible
To Patrice Adcroft, my mentor and great friend

Contents

Acknowledgments

A book of this magnitude and scope cannot be executed without the help of legions of unsung heroes behind the scenes. First and foremost, my incisive editor at *Discover*, Pamela Weintraub, who consistently pushes me to do my best work, assigned me the story that was the original springboard for this book and supportively midwifed this project throughout its lengthy gestation. No writer could have a better champion than my agent, Alice Fried Martell, whose fierce and unflagging enthusiasm, energy, and encouragement have buoyed me through even severe bouts of self-doubt. This book would not have happened without her. My capable assistant, Angelique Robinson, functioned as a combination researcher/therapist, prodding me when I procrastinated, talking me off the ledge when I became overwhelmed, and stopping me from falling down the rabbit hole of minutiae.

Every author should have the exemplary experience I did working with the top-notch team at Rodale. Colin Dickerman, the acquiring editor at Rodale, was among the first to recognize this book's potential. His graceful and witty successor, Alexandra Postman, who came on board midstream, deftly guided this project to completion and lassoed my sometimes-flaccid prose into a taut and seamless narrative. Thanks, too, to Amy King for her sharp design; to Nancy Elgin for all her hard work ensuring my prose made sense, was accurate, and flowed smoothly; and to Nancy Bailey for keeping us on track every step of the way.

Dozens of sources who gave generously of their time have been acknowledged elsewhere. But a handful merit special recognition. Initially, research conducted by Dr. Kim Knowlton of the Natural Resources Defense Council into how climate change was driving the spread of dengue fever prompted me to delve more deeply in this area and ultimately inspired *Fevered*. The CDC's Dr. George Luber's insights were invaluable and our lengthy conversations guided how I structured the book. Down Under, the Australians more than lived up to their reputation as peerless hosts. Public affairs officer extraordinaire Liese Coulter, who was at CSIRO but is now with NCCARF in Brisbane, brought me up to speed on the key issues facing Australia at a time when the learning curve seemed vertical. Similarly, Dr. Chris Cocklin of James Cook University patiently spent many hours giving me his perspective and directed me to key experts who were integral to my research. Elizabeth Hanna of Australia National University not only shared her expertise, but she and her husband, Fergus, also graciously spent a couple of days showing me around Canberra.

I'm also surrounded by a network of supportive family and friends who make me laugh and keep me on an even keel—they are too numerous to mention. However, special kudos to my longtime pal, Edward Silver, who came up with the book's title. On the home front, my husband, Bill Roberts, and my stepson, Nathan, make my life complete and are daily reminders as to why this book is so important.

INTRODUCTION

When the Weather Goes Haywire

"It was the most beautiful day. My husband, Bryce, woke me up before sunup to look at the most beautiful sunrise you ever saw. I was grumpy because it was Sunday morning and we usually slept late on Sunday. But I got up and went out and there were the most beautiful streaks going up from the horizon, way up in the sky, just all colors, just like the northern lights."

That's how Bernice Jackson remembered that fateful Palm Sunday in April 1935 in an oral history interview she gave half a century later to the Oklahoma Historical Society. She was a young farm wife then, living with her husband and their four-year-old son, John, in a clapboard two-story house, raising cows, pigs, and chickens and growing wheat and tomatoes on a hardscrabble patch of the Oklahoma panhandle. Weeks of relentless dust storms—almost 50 in the previous three months—had howled across a parched landscape baked by years of drought and record-breaking heat that often soared into the triple digits; these storms had already blown out five million acres of farmland and destroyed most of the wheat crop in Kansas, Nebraska, and Oklahoma. Parts of Kansas had seen barely a dozen dust-free days since the start of the year, and some areas had suffered dust storms every single day in March. But suddenly, on that cloudless morning, the wind stopped, and an eerie stillness settled into the air. The day dawned sunny and clear, with seemingly endless blue skies and temperatures that climbed into the mid-80s by early afternoon. "Bryce," Bernice told her husband, "maybe that's an omen that our dust storms are over and it's going to rain."

Across hundreds of miles of prairie dotted with hamlets that often contained no more than a post office, a general store, and a church, people streamed outside into the bright, balmy day—as if emerging from a deep hibernation—to tackle the chores they had delayed all spring because of the storms: repairing broken fences and henhouses, planting gardens, sweeping out barns covered in a layer of siltlike dust, attending church services, and driving to visit their neighbors.

Bryce made a deal with Bernice: If she'd help him milk the cows, he'd help her clean up the house. There was dirt everywhere from the dusters, as the storms were called—all over the beds, the floors, the windowsills. Even their pillows bore faint dusty outlines of where their heads had laid. The couple spent most of the day sweeping and scrubbing down the windows, filling up two wheelbarrows with dirt. Then Bernice mopped the linoleum floors in the kitchen and bedrooms until they gleamed. Bryce insisted it was just a foolish waste of time. "Tomorrow the dirt's going to come again," he chided. But it made her proud that the house looked so nice, Bernice recalled, and she was grateful for having one day when they could eat a meal without tasting grit in their teeth. When they were finished, they sat out on the back porch basking in the afternoon sunshine—until they noticed rolling black clouds on the far horizon.

"You and John run in the house quick," Bryce told her, his face suddenly ashen. "I'll go turn the windmill off."

The storm came up so swiftly that Bernice and her young son barely had time to find shelter. As they climbed the stairs, they were surrounded by dust so thick it was black as midnight. "I couldn't see John, but I was holding his hand and boy, did I hang on to that little hand," she said. They groped their way to a chair in the kitchen where they sat wordlessly—Bernice clutching her son tightly on her lap—while the storm rattled the windows with such fury that it felt like the end of the world. Bryce eventually made his way in from the barn, coughing and choking from the dust. The family huddled together in the kitchen, waiting for the storm to pass, shivering because the temperature had dropped by more than 40°F. "Slowly, it got light and we could see the outline of the window," recalled Bernice, who felt lucky to have emerged from the storm unscathed. "But it was quite a while before we could see the barns outside."

No one who had been out gulping the fresh, clean air and enjoying the endless horizons and shirtsleeve weather that morning had any inkling that what seemed like the dawn of a new day was only a temporary reprieve, that a deadly soil tsunami was quietly gathering strength and would soon become the mother of all dusters on a day that is remembered as Black Sunday. In fact, the previous afternoon, a severe storm front had been building 800 miles to the north along the Canadian border, where a warm high pressure system that had been squatting over the Dakotas collided with a cold front from the Arctic, producing blizzards with gale-force winds, frigid temperatures, and heavy snow in eastern Montana. As the cold front marched steadily southward across the High Plains and the Texas Panhandle, the churning winds collected even greater force, kicking up a black wall of soil more than 10,000 feet high and containing 300,000 tons of dirt—twice as much soil as had been excavated to create the Panama Canal—and generating enough static electricity to power New York City.

The storm creeping across the Great Plains would start as a faint roar in the distance. Birds would grow agitated, nervously fluttering and chattering. Cattle tied up in barns bellowed in fear, and rabbits frantically galloped across the prairie. By the time the roar reached its deafening crescendo, whole ranches were buried under a rolling wall of black dust. But in many areas, the brutal storm arrived with little warning, leaving many stranded outdoors with no protection against the elements. Hundreds of people were buried alive, and countless heads of livestock choked to death. When farmers later carved up the carcasses for food, they discovered the animals' insides were packed with mud. Trains derailed and cars stalled on the sides of roads, their engines clogged with dirt and their occupants slowly suffocating.

"Both of us wore respirator masks to make breathing easier, but the fine sand permeates everything," noted Frank McNaughton, a reporter speeding across Colorado with a postal service mail carrier while they tried to outrace the storm, in a United Press dispatch. "It fills the hair and eyes and mouth with grit and makes the lungs feel as though they had been sandpapered inside. The dust filled the inside of the closed car. At times I could see only four feet ahead."

Black Sunday, the occasion of the worst dust storm in our nation's history, provides a grim snapshot of the devastation of the 1930s Dust Bowl and what happens when the weather goes haywire in a landscape that's been drastically altered by human development. Drought and destructive agricultural practices that had eroded the soil set the stage for the worst environmental disaster in American history. Just a scant few inches underneath the fertile soil that extends over the hundreds of thousands of square miles of the Great Plains is sand. These vast expanses of land were once semiarid grasslands only lightly grazed by buffalo and traveled by the Native Americans who subsisted on them for thousands of years. "The flattest, driest, most wind-raked, least arable part of the United States was transformed by government incentive, private showmanship, and human desire from the Great American Desert into Eden with a haircut," noted Timothy Egan in *The Worst Hard Time*, the seminal history of the Dust Bowl that won the National Book Award in 2006. "Settlement was a dare, on a grand scale, to see if people could defy common sense."

This environmental devastation and its human aftermath were famously captured in John Steinbeck's *The Grapes of Wrath*, and ever since I read his wrenching masterwork in high school, I had felt a profound empathy for hardworking families like the Jacksons, whose dreams of a better life were crushed by the seemingly capricious whims of nature. These events unfolded only a little more than a decade before I was born, but as a teenager raised in the heady 1960s, when the glamorous young Kennedys lived in the White House and we were launching men into space, the world seemed to brim with limitless possibilities; this calamity seemed as far removed from my comfortable middle-class existence in suburban New York as the primitive pioneer life of the Old West. It never occurred to me that catastrophes of this magnitude could happen in such a technologically advanced society—or that our profligate exploitation of this land had as much to do with creating the Dust Bowl disaster as the drought did.

But as I began to research this book and initiated conversations with dozens of climate scientists, geologists, and atmospheric physicists, I learned that the Midwest and the Southern Plains are likely to tip into desert once again as temperatures continue their inexorable climb. Combing through the gripping and poignant oral histories of the survivors of the 1930s' Dust Bowl, it hit me that the bleak physical, psychological, and emotional assaults on their daily lives might offer a glimpse of what our future could hold. These stories also provide a disturbing preview of the impending deterioration of our physical health amid the growing deprivations, changes in ecosystems, and increasingly harsh weather that result from the careless dumping of billions of tons of greenhouse gases into the atmosphere.

Even though these changes are happening quickly, we are woefully unprepared and seemingly unconcerned. The effect of a warming planet on our health is "a threat that's been completely neglected, marginalized, and ignored by the global health community and by policy makers," as a team of British scientists recently concluded in a landmark report in the venerable medical journal *The Lancet*. "Yet in terms of our well-being, in terms of our survival over the next 100 years, it is absolutely the top public health issue that we should be talking about."

THE NEW DUST BOWL?

The failure to see what's right in front of us is nothing new. In the 1920s, just like today, few heeded the warnings because the country was gripped by a speculative frenzy. Although the Great Plains wasn't suited to farming, an unusual stretch of wet weather, new technologies that produced more efficient tractors and combines, and inflated grain prices triggered a land boom that resulted in the plowing of more than 100 million acres, up from a relatively scant 12 million acres a few decades before. Thousands of settlers, many of them immigrants from Germany, Ireland, Poland, Russia, and the Slavic countries, lured by the promise of the American dream, were encouraged by land promoters, railroad companies, and state governments to till the Southern Plains and engage in water-intensive European-style agriculture.

When it rained, livestock flourished, crops were abundant, and farmers celebrated record-breaking harvests. Throughout the 1920s, the prairie was a beacon of prosperity in an America that would be ravaged by economic calamity in the wake of the 1929 stock market crash that led to the Great Depression. But when drought struck in 1931, no one was prepared because they had naïvely thought the good days would last forever. "Rain follows the plow" was the refrain, a bit of magical thinking espousing that the very process of cultivating the land would somehow change the climate and quench the earth. Crop prices crashed as the Great Depression deepened. Emaciated cattle died in the fields.

Some who chose to hang on crammed every cranny in their houses with wet cloths, sheets, and gunnysacks in futile attempts to keep out the dust. Even on clear days, the air was so dry that just breathing would sear the lungs, forcing people to wrap dampened bandannas over their mouths and noses and coat their nostrils with petroleum jelly if they dared to venture outside. In that period, people didn't have the social safety nets that we take for granted today, leaving them to the mercy of an uncertain fate. Many were isolated by the storms that crippled cars and machinery, marooned for days on end without food, barely surviving on the ragged edges of poverty. Some were driven mad by the never-ending dust.

By 1935, inhabitants began to abandon the Plains in what became the largest mass migration in US history. At the end of the decade, on the eve of World War II, more than two million people had been uprooted—including 85 percent of the population of Oklahoma—and 500,000 more were left homeless. Deserted farmhouses covered the landscape, and the institutions that formed the delicate fabric of communities—schools, churches, banks, and businesses—were shuttered, leaving behind ghost towns. Many of those who left found themselves living in squalid migrant camps in California's Central Valley, picking crops dawn to dusk under the desert sun when they could scrounge work, stigmatized by the locals as "Okies" and white trash. Some families never climbed out of poverty. Those who remained on the Plains were demoralized as they watched their livelihoods being blown away by the storms.

Aside from the profound psychological toll, there also were serious health consequences from living in a dust-choked environment where summer temperatures soared to 120°F and air-conditioning was exceedingly rare. Simply venturing outdoors could prove fatal. Millions suffered from malnutrition and even starvation, and many subsisted on pickled tumbleweed, yucca roots, and roadkill. Even the dust itself was lethal: The churning winds milled the soil into an extremely fine particulate with a high silica content, which scratched throats and eyes and penetrated deep into the human body. The dust collected in tiny pockets inside the lungs, causing a potentially lethal condition known as "dust pneumonia" or the "brown plague" that was similar to the black lung that developed in coal miners. No definitive public health records were kept during that period, but experts estimate that as many as 7,000 people suffocated from the dust and thousands more were permanently incapacitated, condemned to a lifetime of hacking coughs and respiratory difficulty—and, in some cases, early death.

Will we see dust bowl conditions again? Absolutely. By the end of this century, many experts told me, we can expect the midsection of the country to be gripped by extreme droughts and baked by 90°F days for more than half the year. But at a certain point, as one meteorologist pointed out during the nation's most recent dry spell, we will no longer call them droughts because the weather patterns will have permanently shifted and the land will simply become desert. It's a scenario that doesn't seem so farfetched after witnessing the events of 2012, the warmest year in recorded history, which may finally have triggered a seismic shift in the public's awareness of climate change. As the cover of the November 1 *Bloomberg Businessweek* starkly noted after Hurricane Sandy turned much of the northeastern United States into a gigantic disaster zone: "It's Global Warming, Stupid."

Starting in early March 2012, much of the nation sweltered under triple-digit temperatures in a series of unprecedented heat waves—3,282 daily temperature records were broken in the month of June alone, and July was the hottest month since record keeping had begun in 1895. A severe wind storm—a type called a "derecho," which sweeps across hundreds of miles, with wind gusts of at least 58 miles per hour, lasting more than six hours, and is often linked with thunderstorms or showers—cut off electrical power to 3.7 million people in Ohio, Pennsylvania, West Virginia, and Virginia for up to a week. The exceptionally dry conditions ignited raging wildfires across the western United States—Utah, Nebraska, Wyoming, Montana, New Mexico, Arizona, and Colorado—that incinerated more than seven million acres. The heat also contributed to the record-breaking drought, the worst since Dwight Eisenhower was president, that engulfed 80 percent of the continental United States by mid-July, affected 165 million Americans, and decimated 65 percent of cattle production and 75 percent of the corn crop.

At the end of October came Hurricane Sandy, which even normally reserved scientists described as "historic" and "unprecedented." The sheer kinetic energy generated by the storm surge and the destructive potential of the waves in its wake reached 5.8 on the National Oceanic and Atmospheric Administration's 0-to-6 scale, the highest ever measured. Exceptionally high ocean temperatures brought this very-late-season Atlantic hurricane barreling up the East Coast to crash into a cold front that was storming down from Canada. The front's frigid air, according to researchers at

Cornell University, was fueled by the unprecedented September melting of Arctic ice, which had shriveled to 1.32 million square miles, the smallest ever recorded, and less than half the area it had occupied only four decades ago. The collision of these two weather systems turbocharged Sandy and transformed it into a "Frankenstorm" that stretched about 850 miles—greater than the distance from New York to Atlanta—and caused historic destruction and catastrophic flooding in the nation's most populous regions, including many communities near where I grew up in New York City's northern Queens. I watched with an overwhelming sense of dread as the devastation climatologists had been predicting for years unfolded before my eyes: At least 110 people were killed and thousands lost their homes; enormous swaths of the electrical grid failed, leaving millions—including much of lower Manhattan—without power for days, and some for weeks; New York's entire subway system was shuttered for days; LaGuardia Airport was submerged; and many of New Jersey's iconic beachfront resort towns were turned into piles of kindling.

As the weeks wore on, many people in the worst-hit areas of New York City, such as Far Rockaway, a poor neighborhood on a peninsula of land that protects Brooklyn and Queens from the open waters of the Atlantic and has fortress-like public housing towers built along the beachfront, were forced to remain in their demolished and mold-infected homes in a scenario that echoed what had happened in New Orleans after Katrina. News accounts about how these stranded residents were sickened by serious respiratory infections and developed what came to be known as "Rockaway cough" brought to life with stunning clarity the warnings I heard repeatedly from the health experts: As the planet gets hotter, we'll live sicker and die quicker.

WHY THE PLANET IS HEATING UP

For more than 100 years, scientists had cautioned that burning fossil fuels like coal and oil would cause global warming because it adds enormous amounts of carbon dioxide to the atmosphere. Carbon dioxide is considered a "greenhouse gas" because it creates a hothouse effect by absorbing infrared radiation from the sun, inhibiting the planet's natural cooling mechanisms. While greenhouse gases are normally present in the environment—plants use CO_2 for photosynthesis, for instance, and we exhale it every time we breathe— we've released tons more into the atmosphere since coal came into widespread use in the early 19th century, at the dawn of the Industrial Revolution. With more carbon-spewing vehicles and factories constantly coming online to accommodate population growth, carbon emissions continue to climb. By 2011, annual global carbon dioxide emissions had reached 31.6 gigatons, an increase of 3.2 percent over the previous year, and there seems to be no end in sight. To quantify the magnitude of all that carbon, a gigaton is one billion tons, equivalent to about twice the mass of all seven billion people on earth. *Twice.* This means that 31.6 gigatons is more than 60 times the aggregate weight of every single person on this planet. That figure is expected to rise by about 3 percent annually as the population swells and more people around the globe enter the middle class and consume more energy.

Indisputably, the planet is heating up. The clearest depiction I have heard of the environmental consequences of dumping unnatural amounts of carbon dioxide into the atmosphere came from Steven Sherwood, an atmospheric physicist I met in Cairns, Australia, where I spent nearly a month taking a hard look at the front lines of climate change. The American-born Sherwood has a glossy résumé: He studied physics at MIT, did his doctoral work at the Scripps Institution of Oceanography at UC San Diego, and spent nearly eight years teaching in the geology department at Yale University before taking over as codirector of the Climate Change Research Center at the University of New South Wales in Sydney. Now in his 40s, he looks like a stereotypical distracted scientist, with a nimbus of unruly dark-blond hair and wire-frame glasses, and radiates the prickly brilliance of someone who does not suffer fools gladly.

"We've always had weird weather," he told me, and it's hard to blame each event on climate change. "But we're really blowing the lid off on CO_2 emissions," he warned, "and it's looking more likely that temperatures will rise by at least four degrees Farenheit by the end of the century, which could make the earth hotter than it's been since the dawn of civilization." The fact is, he pointed out, the planet is heating up—on average, it's 1.4°F warmer than it was 100 years ago—and it's probably hotter than it's been in at least 1,000 years. Scientists have taken into account several environmental indicators to make this judgment, he added, ticking off each factor: the sea ice is retreating; permafrost and glaciers are melting; species are migrating northward to find more hospitable climates; tree-ring data culled from the last millennium show temperatures are climbing; and spring thaws occur a week earlier and winter freezes commence a week later than they did 50 years ago.

But couldn't this change be just a natural variation? Here again, Sherwood said that analysts have run computer models of every possible permutation and factored in every conceivable confounding variable and arrived at the same conclusion: Since 1970, the warming has been too rapid and too widespread—glaciers in places like Alaska, Montana, western Canada, and eastern Russia and polar ice shelves formed thousands of years ago are now shrinking swiftly—to be caused by normal climatic shifts, which usually occur over the span of several centuries. Plus, numerous studies show that the last two decades of the 20th century were the hottest in 400 years and perhaps the warmest in several millennia.

Rapid ocean acidification—which increased by 30 percent in the past century—is another tip-off. The oceans are the world's carbon sink, absorbing about 50 times more CO_2 than the air does. But carbon dioxide forms carbonic acid when it's dissolved in water. As a consequence, rising CO_2 emissions are fueling the growing acidity of the oceans, which is killing seafood species, coral reefs, and organisms at the foundation of the ocean food chain. By 2050, if carbon emissions continue at current rates, the alkalinity of the ocean will be lower than at any time in the last 20 million years, a change that is occurring 100 times faster than at any time since Earth was formed.

But are burning fossil fuels and the rising level of carbon dioxide the culprits behind the hotter temperatures? Scientists have used known physical laws to prove that industrialization, deforestation, and pollution have supercharged the concentrations of greenhouses gases such as carbon dioxide, methane, and nitrous oxide in the upper atmosphere. These gases absorb extra solar radiation and then release that excess heat into the lower atmosphere, inhibiting planetary cooling and creating a hothouse environment under the carbon canopy that amplifies temperatures on Earth's surface. And this has all happened since the Industrial Revolution. "We now have very accurate histories showing little change [in CO_2 levels] for millennia, then a meteoric rise after coal came into wide use in the early 1800s," Sherwood has stated. "The key thing is that we are hitting the climate system harder than any known natural forces . . . and the recent warming is right in line with long-standing predictions."

In the coming decades, as I discovered when I talked to climate experts and public health officials from all over the world—the United Kingdom, Australia, Russia, and the Netherlands, and North American cities as diverse as New Orleans, Philadelphia, Chicago, Miami, Los Angeles, New York, Seattle, and Vancouver, Canada—we'll be living with the effects that higher temperatures have on ecosystems: higher levels of ozone pollution in the air we breathe; more uncontrolled outbreaks of deadly infectious diseases as mosquitoes migrate to newly warm habitats; and more extreme weather events. Hot air holds more water, so we will have more torrential rains, more ferocious hurricanes, and, conversely, more dry spells as a result of heat-induced changes in rainfall patterns. Rising temperatures could trigger pestilence, drought-induced food shortages, raging firestorms, massive migrations, political instability, and wars, even the return of the bubonic plague, the Black Death that killed more than 25 million people in the Middle Ages. In the near future, millions might perish and millions more might be sickened by the litany of medical conditions caused or exacerbated by living in a rapidly warming world: heart disease, asthma, severe respiratory infections, heatstroke, and suicidal despair.

Then there is the cascade of collateral damages: the debilitating injuries and deaths that come with increasingly violent and more frequent hurricanes, floods, and fires, and the chronic illnesses exacerbated by being left untreated for lack of medical care after weather-related calamities. Think of New Orleans, whose public health system was flattened by the devastation wrought by Hurricane Katrina. The storm destroyed more than a dozen hospitals and uprooted thousands of doctors, causing virtually all residents to lose access to their usual health care providers. The scale and scope of the damage from Katrina have made rebuilding Louisiana's health care safety net a massive undertaking.

Charity Hospital, a beacon for the city's poor and uninsured for more than two centuries, remains shuttered. Until December 2008, more than three years after Katrina made landfall, there was only one Level 1 trauma center to service a city of nearly half a million. That meant people with severe brain injuries—for whom precious minutes can mean the difference between recovery and a lifetime of disability or death—sometimes had to wait an hour or more before they could receive lifesaving care. Massive numbers of doctors and nurses have left the area, and many of the remaining hospitals are still not operating at pre-Katrina levels, according to a recent report by the Urban Institute, a Washington, DC, public policy think tank. Today, the average life expectancy of an African American in New Orleans is about 69 years, nearly as low as it is in North Korea.

Even cities with elaborate disaster management systems are unprepared for the shattering impact of our harrowing new weather. In Joplin, Missouri, which lies squarely in Tornado Alley, tornado drills are as much a part of the city's life as Fourth of July barbecues. In May 2011, Joplin's tornado warning system gave residents 24 minutes' notice that a twister was on its way. Yet they were powerless in the face of the storm's ferocity: More than 150 people died, including 4 at Saint John's Regional Medical Center, where the roof was ripped off and the building was left an utter shambles. Afterward, eight tornado victims with multiple injuries fell ill from a rare and deadly fungal infection that enters the body through a puncture wound or by breathing in the mold's spores. Normally, the infection can be prevented by thoroughly disinfecting injuries. But in the confusing aftermath of the disaster, when doctors and nurses at Saint John's were treating hundreds of injured residents in makeshift clinics, health officials believe the wounds may not have been adequately treated.

When Superstorm Sandy hit the East Coast, major hospitals like New York University's Langone Medical Center and Bellevue lost power, and hundreds of gravely ill patients had to be laboriously transported down several flights of darkened stairwells and evacuated into waiting ambulances in the midst of ferocious winds and driving rains. These two facilities, along with the Manhattan Veterans Affairs Medical Center and Coney Island Hospital, didn't fully reopen for months, forcing patients who needed to be admitted to find medical care at other area hospitals that were already stretched beyond capacity. Even weeks later, long after most of the city had returned to some semblance of normalcy, teams of federal disaster recovery workers discovered dozens of elderly and infirm residents living in squalor, trapped in high-rise towers. Unreliable electrical systems in many buildings were devastated by the flooding, crippling elevator service and leaving nearly 40,000 apartments without heat and hot water. Volunteers found stranded people sitting in wheelchairs with no way of escape, diabetics who lacked insulin, and others who didn't have food or were staying warm by turning on gas stoves that emitted deadly carbon monoxide fumes.

In the absence of meaningful mitigation and adaptive strategies, we are on the cusp of a terrifying and increasingly unhealthy future. "Climate change will cause a worsening of the common health problems we already see," Dr. John Balbus told me in a series of interviews. Balbus is a physician and a senior policy analyst who leads the climate change and health effort at the National Institute for Environmental Health Sciences. "We're going to see incremental changes in the next 5 or 10 years, but that might not compare to what we're going to see in the next 30 or 40 years," he said. "Scientists didn't want to be alarmist, but they have systematically underestimated the threat."

ONE DEGREE

One degree. When I talked to meteorologist Siegfried Schubert, I was astonished to find that's all it took to cause the 1930s Dust Bowl. Just a 1°F change in the surface temperatures of the oceans cut off the pipeline of moisture that normally travels north from the Gulf of Mexico and triggered the long dry spell, according to a study he did more than 60 years later, in 2004, at NASA's Goddard Space Flight Center in Maryland. Granted, agricultural techniques, such as deep plowing that removed the top layer of fertile dirt and grasses that anchored the soil and a failure to rotate crops, which depleted vital nutrients, made the fields more vulnerable to wind erosion and created the conditions for the dust storms. The European immigrants who moved to the Great Plains unwittingly adopted cultivation methods that were better suited to the wetter climes of their countries of origin than to the semiarid prairies. Unfortunately, the largest wave of settlers had arrived at the turn of the last century, during an unusually wet period on the Plains, so they expected this weather to last forever. And as long as it continued to rain, farmers escaped the consequences of their misguided practices. "The 1930s drought was a major climatic event on such a great, grand scale," Schubert told me, elucidating why he had conducted this research. "It's something that has to be explained, especially if we are to make progress in understanding future droughts in relation to global warming."

Using sea surface temperatures collected over the past 100 years by ships at sea, Schubert fed the data into a computer model to run a simulation of the 20th century's climate patterns. He discovered that a slight oscillation, less than 1°F, in the surface temperatures of the oceans—a cooler than normal tropical Pacific Ocean surface temperature combined with a slightly warmer tropical Atlantic Ocean temperature—created the atmospheric conditions that caused the drought. The temperature change in the surface water of these two great oceans disrupted the course of a band of low-lying air that is responsible for bringing rain to the Plains. This ribbon of fast-moving air near Earth's surface typically flows westward over the Gulf of Mexico, where it sucks up moisture and then swings northward, dumping rain on the Great Plains. But in 1932, the warmer Atlantic heated the air, making it lighter and less dense; as it rose, it allowed cooler, higher-pressure air to sweep in off the continental United States, pushing the wet air eastward over the Gulf. Cooler temperatures over the Pacific also interfered with this cycle, strengthening the summertime high-pressure system that sits over the Great Plains and driving the stream of air farther south than normal. The air that returned was even drier. The interplay of these two currents reinforced them, said Schubert, and these conditions persisted for at least six years, which created the perfect storm that left the Midwest and the Great Plains dry as pavement.

The dust itself also enhanced and prolonged the drought, establishing a feedback loop. Dust reflects sunlight and radiation, according to Schubert. Because the dust layer blocked the sun's rays, it left the ocean surface cooler. Plus, because the soil was dry, less moisture evaporated into the air, which resulted in even less rainfall and another negative feedback loop that probably doubled the effect of the drought, Schubert said. Little wonder that in the summer of 1934, drought conditions covered almost the entire country. Schubert told me that he was as surprised as I was that such a small temperature variation could cause a drought of this enormity and duration. But he pointed out that this just underscores the climate's sensitivity and vulnerability to small movements in the thermometer—even a few 10ths of a degree can make a big difference.

Most of us who live in a temperate climate are accustomed to wide temperature swings over the course of a year, from freezing-cold winters to sweltering summers. The notion that temperatures will gradually increase an average of a few degrees doesn't seem like much of a threat. That's why Schubert's research is so unnerving: It illustrates the fact that seemingly minor temperature changes can trigger a chain reaction of alterations in delicately calibrated ecosystems and shifts in weather patterns with long-lasting consequences.

While there have been some notable fluctuations, for the past 12,000 years we've enjoyed a relatively stable climate that has allowed civilization to flourish. But we are now on the threshold of transformative changes in the weather. Although international public support is finally building to take action against global warming, there is still a pervasive—and falsely comforting—belief that climate change will happen slowly, that the globe will heat up uniformly, and that the predicted devastation won't kick in until long after the baby boom generation has died of old age. The developing world—Africa, Asia, and South America—will bear the brunt of the toxic legacy of wealthier nations' addiction to fossil fuels. But even relatively affluent Americans won't be observing this seismic shift from a safe, insulated distance. Disruptive changes will hit hard here, too, and sooner than we think. "People think we can just adapt our way out of this," George Luber, an epidemiologist and the associate director for global climate change at the Centers for Disease Control and Prevention (CDC) in Atlanta, told me in a series of interviews. "But this isn't about gradual changes. This is about climate viability that will drive extreme events."

Already, the fingerprints of climate change are etched across the landscape in the catalog of freak weather patterns occurring across the United States. These confirm that Earth is warming at a swifter pace than even the direst forecasts predicted a few short years ago. "With respect to sea-level rise, with respect to temperature changes, with respect to carbon emissions, and in just about every case, the changes have occurred either at the upper end of the projections or even above the range of the projections," said Michael E. Mann, a climate modeler at Pennsylvania State University, in a recent interview with *Scientific American*. These days, nightly newscasts, to borrow a phrase from Al Gore, often "resemble a nature hike through the Book of Revelation."

Since John F. Kennedy was president, the United States has heated up more than 2°F, a change greater than the warming average for the whole planet. Winters are now shorter and warmer than they were 30 years ago, with the largest temperature rises—more than 7°F—measured in the Midwest and northern Great Plains. Even if we move at warp speed to reverse this trend by drastically cutting carbon emissions—which clearly is not happening—temperatures will continue to climb because of the heat-trapping carbon dioxide that's already been dumped into the environment. Carbon dioxide lingers in the atmosphere for centuries, while oceans absorb the heat caused by warming and release it back into the air for hundreds of years. "Twenty percent of the global-warming pollution we spew into the sky each day will still be there 20,000 years from now," noted Al Gore in a 2011 essay in *Rolling Stone*, which means that we're condemning "our children and all future generations to struggle with ecological curses for several millennia to come."

Over the next century, the thermostat will climb another 2°F to 11°F degrees on average, a range that is contingent upon what we do to reduce greenhouse gases, according to the projections from numerous government studies done both in the United States and abroad. The lower end of that prediction is virtually guaranteed because of the accumulated CO_2 already in the atmosphere, according to a 2010 report from the National Research Council, which is one of the four National Academies that advise the country on scientific matters. This translates to more extremely hot summers in every part of the country. When the amount of carbon dioxide in the atmosphere climbs from the current 393 parts per million (ppm)—up from about 298 ppm in 1900—to 600 to 700 ppm and beyond by the end of this century, as climate modeling scenarios now anticipate, we'll be living under a carbon blanket far worse than the suffocating cloud of smog that envelops today's most polluted megalopolises—think Beijing and Shanghai. Conditions such as these will not only make breathing a chore, they will change the climate in ways so profound and causing such vast and far-reaching disruptions to ecosystems, rainfall, and water supplies that the world will be virtually unrecognizable.

Based on current projections by the National Center for Atmospheric Research, Earth's most populated areas—a huge expanse of land extending from northern Canada to the southern tip of South America, parts of Asia, and most of Australia and Africa—will be parched by drought by the century's end, drying up surface water and killing crops that hundreds of millions depend upon for survival. "If the projections come even close to being realized," Aiguo Dai, the climatologist who calculated them, told *Rolling Stone* in 2011, "the consequences for society worldwide will be enormous."

A 5°F rise is at the very outer edge of what we may be able to manage— anything higher than that raises serious questions about our survival as a species. "Human-made climate change is almost certainly going to be the greatest moral issue of this century," said climatologist James Hansen in a recent speech presented to a standing-room-only crowd at UCLA. The chief climate scientist at NASA's Goddard Institute for Space Studies in Manhattan, Hansen is a mild-mannered septuagenarian with thinning gray hair. Despite his small-town Iowa Corn Belt roots, the astrophysicist has become an environmental rock star. He's been warning about the serious impact rising temperatures will have on the climate for nearly a quarter century. He even chained himself to buildings and got arrested during demonstrations outside the White House protesting the George W. Bush administration's policies. In 1988, in testimony before the US Senate, he stated unequivocally that global warming caused by human activities had already begun. "There is a huge gap between what is understood about global climate change by the relevant scientific community and what is known by the public," Hansen told his audience at UCLA. "The climate system has enormous inertia and doesn't change that quickly. We have only felt a fraction of the climate change from the gases already in the atmosphere. Still more is in the pipeline. It's hard for people to recognize that we have a planetary emergency."

Many scientists now believe we're on the verge of passing several tipping points that could unleash unstoppable forces. Melting permafrost in Siberia could belch billions of tons of methane—a greenhouse gas 20 to 70 times more potent than carbon dioxide—into the atmosphere and set off feedback loops that reinforce and amplify runaway climate change. As a consequence, many experts say climate change will be abrupt and cataclysmic, and that some areas, especially here in the United States, will be more impacted than others. "I can't rule out that it's already too late," said Stephen Schneider in what turned out to be one of the last interviews the Stanford University climatologist gave before his unexpected death at age 65. Fearless and outspoken, Schneider, a former MacArthur "Genius" fellow, had been trying to get the world's attention for 30 years, and he no longer had the patience for sugarcoating the truth to make it politically palatable. "The trouble with tipping points is that you don't know if you've crossed them for 50 to 100 years because of the time frame of the system," he told me. "We're looking at an amazing fiduciary responsibility for our behavior in this and the next generation because it will predict the conditions in this world for the next 150 generations. But we won't know until after it is too late."

A Threat Multiplier

Climate change is a "threat multiplier," in military parlance. That means the existing hazards become amplified in what sociologist Christian Parenti calls a "catastrophic convergence." Climate change is not just about melting icebergs, storm surges, floods, and droughts, according to Parenti. A warmer world will be more dangerous and unstable, and the damage caused by climbing temperatures will be a catalyst for escalating religious conflicts, civil wars, failed states, mass migrations, and border militarization to halt the influx of desperate refugees as the planet deteriorates.

Newly industrialized developing nations such as Brazil, China, India, Indonesia, Mexico, and Turkey are rapidly creating a huge middle class that will demand more goods and create more pollution, putting additional pressures on a global ecosystem already buckling under the weight of human consumption. Yet vast stretches of the world could become virtually uninhabitable, forcing the exploding population—expected to reach nine billion by 2050—to squeeze into ever smaller patches of livable land. As we start to feel the full effects of climate change, the rule of thumb is that every 1°C rise in temperature—a little less than 2°F—flattens crop yields by 10 percent, according to a widely cited 2004 Chinese study in the *Proceedings of the National Academy of Sciences*. Higher temperatures halt photosynthesis, prevent pollination, and lead to crop dehydration. For an example, look no farther than Russia, where the extended heat wave and drought in 2010 cut their grain harvest by 40 percent. How will we grow more food to feed all those extra people on a planet with more frequent droughts, floods, and heat waves? Food prices could double, pushing billions into starvation. Unless we begin reducing carbon emissions, Parenti warned in a recent interview with *Foreign Policy in Focus*, "we are going to hit all the tipping points climatologically, which will lead to self-compounding climate change and the unleashing of such radical transformations in weather patterns that it will be very hard for civilization to hold on. Radically rising sea levels and the massive desertification of the grain baskets of the world, among other problems, will make it very hard for even the most developed economies to survive."

The World Health Organization (WHO) estimates that, worldwide over the past three decades, 150,000 people a year have died as a result of a warming planet—mainly from increased mortality due to higher rates of malaria, diarrheal diseases, and floods—and that five million cases of illnesses can be attributed to it annually, too. And those figures don't include the tolls from heat waves, crop failures, and other deadly diseases that have increased as the planet warms. Up to five million deaths occur each year from air pollution, hunger, and disease as a result of climate change and emissions from carbon-intensive economies, according to a 2012 report commissioned by the Climate Vulnerable Forum, a consortium of 20 developing countries strongly threatened by a warming world. If current trends continue, 100 million more people will die by the year 2030, the annual death toll will rise to 6 million, and global economic growth will diminish by 3.2 percent a year. This is only the beginning, but this medical meltdown isn't going to happen in some distant future. "We're starting to see irregularities that are difficult to explain," said the CDC's George Luber. "Outbreaks of rare fungal diseases in the Pacific Northwest that were normally confined to the tropics, or harmful algae blooms killing off fish in the Florida Keys that we used to only see in the south Caribbean."

The WHO, in fact, has identified more than 30 new or resurgent diseases in the last three decades, the sort of explosion some experts say has not occurred since the Industrial Revolution brought masses of people together in overcrowded cities and vermin-infested slums rife with infectious diseases. The incidence of dengue fever, long thought eradicated in the United

States and once close to being wiped out in South America, is now climbing in the Western Hemisphere—the number of people hospitalized in the United States with it tripled between 2000 and 2007, according to the CDC—and the species of mosquito that spread dengue have established a firm foothold in the continental United States.

A hotter planet is also promoting the spread of numerous other vector-borne pathogens. Ticks, mice, and other carriers of potentially deadly microbial hitchhikers are surviving milder winters and fanning out across the country into newly suitable habitats, transmitting Rocky Mountain spotted fever, equine encephalitis, Saint Louis encephalitis, and babesiosis, a once uncommon malaria-like infection. Lyme disease has migrated from Connecticut and New York to the Canadian border and westward to the Great Lakes region. The sweltering summer of 2012 saw the largest outbreak of West Nile virus ever in the United States, according to the CDC, with 38 states reporting 1,118 cases, including 41 deaths.

Heat waves like the ones that killed more than 70,000 people across Europe in 2003 and 2005 are projected to become common. In 2010, Russia wilted under its most intense heat wave in 130 years of record keeping, with daily highs in Moscow hitting more than 100°F instead of the normal summer average of 75°F. Severe droughts ignited wildfires in the countryside, smothering Moscow in poisonous smog for six straight days. The combination of unprecedented heat and suffocating haze doubled death rates to an average of 700 people a day, and more than 52,000 overall. Here in the United States, more than 150,000 Americans could die by the end of the century from the excessive heat caused by climate change, according to a 2012 analysis by the Natural Resources Defense Council, with cumulative heat-related fatalities in cities in the nation's midsection like Louisville soaring to 19,000; Detroit is expected to have 17,900, and Cleveland 16,600. Scorching temperatures in the 110°F to 120°F range could become the new norm in many parts of the country, especially the Sunbelt.

Big cities will feel the heat more acutely because of their high concentrations of asphalt, buildings, and pavement, which tend to absorb more heat in the day and radiate less heat into their immediate surroundings at night than rural areas do. Therefore, built-up areas get hotter and stay hotter, creating "urban heat islands" in which temperatures are 5°F to 10°F warmer than surrounding areas. In the not-too-distant future, major metropolises like New York, Chicago, Philadelphia, and Phoenix could become uninhabitable hot zones for months at a stretch, triggering the deaths of thousands.

Allergies and asthma have reached epidemic proportions in industrialized nations. Asthma rates have increased by 50 percent in each decade for the last 40 years, according to a 2008 survey by the World Allergy Organization, and more than 300 million people worldwide now have asthma, while an additional 400 million have allergies. Already, at least 50 million people in the United States have allergies, while asthma affects 1 in 14 American adults and nearly 10 percent of children, making it the leading cause of school absences. The incidences of both respiratory conditions are ticking upward partly because pollution and pollen worsen as the thermometer rises. Rising temperatures have also resulted in earlier and longer pollen seasons. More potent allergens, such as the pollen in ragweed, are being produced in higher quantities because of warmer temperatures and because the air contains higher concentrations of carbon dioxide.

Scientists now know, too, that huge dust storms like the ones that recently blanketed Arizona, northern China, Australia, and other arid areas are also responsible for spreading lethal epidemics around the world. While experts used to believe diseases were mostly transmitted by people or animals, many now maintain that airborne dust clouds can carry viruses like influenza or SARS (severe acute respiratory syndrome) and other potentially harmful bacteria, viruses, and fungal spores over thousands of miles. Storms across the Sahara have been blamed for the spread of fungal meningitis spores, which infect more than 250,000 people a year. Domestically, higher temperatures and more intense storms are linked to coccidioidomycosis—"Valley fever"—a sometimes fatal disease infecting more than 200,000 Americans annually that is contracted by breathing in a fungus found in soil in the southwestern United States, Central America, and South America. In the past decade, incidence of this illness has quadrupled in the drought-stricken Southwest.

Sociological trends will also exacerbate the spread of disease as the planet heats up. The drought-driven economic collapse occurring in many rural farming communities in developing countries has escalated migration to the world's megacities. The urban shantytowns that sprout up on the fringes of these giant metropolises tend to be filthy and overcrowded, making them breeding grounds for contagions. Increased global travel has stepped up the transmission of tropical infectious diseases to industrialized nations. As a consequence, outbreaks of malaria in megacities in the equatorial belt—Sao Paulo, Manila, Nairobi—are now just a plane ride away from cosmopolitan urban centers like Miami and Los Angeles. The speed with which SARS spread from pig farms in rural China to North America is just one example of how epidemics go global.

There is also the psychological fallout of living through more frequent natural disasters—the breakdown in social cohesion, the lost income, debt, and property damage—that can spill over into mental health problems such as anxiety, depression, post-traumatic stress disorder, substance abuse, domestic violence, and suicide. In the wake of repeated wallops by a series of catastrophic weather events—a decade-long drought, severe cyclones, brushfires, and record floods—Australian researchers found that global warming created a "climate of suffering," and that one in five people felt long-lasting after-effects, including depression and anxiety. By the year 2020, the worldwide financial and societal burdens of mental illnesses will be greater than for any other major disease category, according to the WHO.

In researching this book, I drilled down into each of these phenomena in detail—the heat waves across Europe, the devastation in Australia, the increasingly polluted air in industrial centers, and the fallout from such natural disasters as Hurricane Katrina and Superstorm Sandy. What I discovered is that these repeated assaults could lead to the collapse of our normally well-functioning public health system. Ironically, the 20th century was a period of triumph for modern medicine. Life expectancy in the United States shot up rapidly with the commercial availability of penicillin after World War II, from 47 years in 1900 to 68 in 1950, and steadily increased through the subsequent decades with improved nutrition, better hygiene and sanitation, the development of drugs that helped tame scourges like heart disease and cancer and vaccines that vanquished such ills as polio, and the construction of a vast and vigilant public health network tethered to the mother ship, the CDC. Founded in 1946 to eradicate the malaria that was pandemic in the Midwest and the South, the agency has since led the world in the containment of deadly epidemics. Yet as the planet heats up in the new millennium, the CDC's success may well be reversed.

"Our society would be overwhelmed if we have multiple Katrinas in a given year," Michael McGeehin, former director of the Division of Environmental Hazards and Health Effects for the CDC's National Center for Environmental Health, told me in a recent interview. "When you have Katrina-like flooding in the northern Mississippi River, what happens to the public health system? You go back to the 1890s—we go back 110 years because it is so easy to stress the system to the breaking point where we don't have basic sanitation, uncontaminated food or water, or the ability to control communicable diseases."

WHAT IS TO BE DONE

Initially, I was deeply pessimistic that we could make the needed changes in time to avert catastrophe. Up until quite recently, the Obama administration has paid lip service to the looming threat of global warming, and most Americans are, understandably, preoccupied with economic survival, not carbon emissions. The impact on our health barely rates any media coverage, even though that's the part of a changing climate that will hit us the hardest in the near future. The United States has neglected to support research into the health consequences of climate change, according to a 2009 report by the US National Institute of Environmental Health Sciences; in 2011, we spent a paltry $3 million to study this issue, a federal oversight that researchers warned "is needlessly putting multitudes at risk."

State governments, with a few notable exceptions, have done little to plan for the added burden on the health system. While 33 US states have strategic plans for dealing with climate change, only 5 of those currently include a public health response, according to a 2009 report by the Trust for America's Health, a nonprofit health advocacy group based in Washington, DC. "We only respond after a Katrina-like crisis," Jeffrey Levi, the organization's executive director, told me. "But we have allowed our core public health capacity to diminish over time. Unfortunately, we don't know how well prepared we are until we're tested."

But there is also some good news—more, actually, than I had imagined, which made me more hopeful that our much vaunted American ingenuity and ability to pull together in times of crisis will save us. The choices we make now will determine not only our fate, but also that of generations to come. On a national scale, we need to build a strong public health infrastructure that will be able to cope with the widespread effects of climate change, such as beefed-up surveillance and warning networks for disease outbreaks, heat wave management programs, and a more resilient public health system that won't be flattened by disasters. But at a local level—where the rubber meets the road—some responsible and savvy civic leaders already recognize that the climate is changing. All they have to do is look out the window to see it, one administrator told me. Mayors in cities like Los Angeles, Seattle, New Orleans, Philadelphia, Miami, and New York are helping to lead the way by working in concert with local community groups to identify vulnerabilities like increasingly bad air due to carbon emissions, more intense and frequent storms or droughts, and sea-level rise and the inland intrusion of salt water in places like Miami. They are taking the first tentative steps toward building a greener, healthier future by ensuring residents will have clean water and clean-energy transportation, and strengthened infrastructure that can withstand the floods, wildfires, and disease outbreaks that will intensify as temperatures rise.

Because many of the threats associated with global warming are generally predictable, it is possible to design or adapt buildings and communities to be more resilient. The urban heat island effect in cities, for example, can be offset by increasing green space—rooftop gardens, more public parks—and reducing black pavement. Cities that anticipate being hit with hotter weather, like Chicago and Philadelphia, have put model programs in place to prevent loss of life during heat waves.

Many of the strategies for creating a nonpolluting, clean-energy future can also improve public health. Reducing our dependence on cars should mean that more people walk and cycle, leading to a decrease in obesity. Right now, for instance, roughly 60 percent of Americans don't get enough exercise. By simply taking public transportation, commuters get an average of 19 minutes of added exercise a day—which is roughly two-thirds of the exercise they need daily. Valuable lessons can be learned from eco-friendly

cities like Portland, Oregon, and San Francisco, which have implemented
sustainable development programs designed to cut traffic congestion,
reduce greenhouse gas emissions and gasoline consumption, and encourage
walking, hiking, and biking.

We need look no farther than North America's uncontested leader in
smart growth, Vancouver, Canada, which has held the line on urban sprawl
thanks to farsighted planning. It never built an extensive system of freeways—city fathers considered them but ultimately rejected the idea after a
heated grassroots campaign literally stopped the bulldozers in their tracks—
and instead encouraged the development of high-density, pedestrian-
friendly neighborhoods. Traffic congestion is controlled by offering
alternatives to cars, including buses and light rail, as well as a growing network of bike trails.

But swift and decisive action will be required to solve the medical issues
that are arising as the planet warms, and the United States must take the
lead in implementing a sweeping coordinated international effort to create
a robust infrastructure capable of blunting the impact of global warming on
our health and well-being—a medical Marshall Plan modeled on the massive rescue effort that saved Europe after World War II. Unless steps are
taken to stem the release of greenhouse gases and adaptive strategies for
living on a hotter planet are put into action, billions may die.

That was certainly the judgment of *The Lancet* report's lead author,
Dr. Anthony Costello, when I talked to him at length over the phone from
his office at University College in London, where he heads the UCL Institute
for Global Health (formerly the Center for International Health and Development). Pediatrics is a "calling" for the affable British physician, who's
spent the past three decades in some of the world's most destitute and
remote regions in Bangladesh, India, Malawi, and Nepal. He's dedicated his
life to improving the health of expectant mothers and their newborns, and
reducing what he calls the "silent emergency" of infant and maternal mortality in impoverished nations. Costello admitted to me that he hadn't
thought climate change was much of an issue until he looked into its links
to natural catastrophes like floods, cyclones, and famines, all of which have
serious effects on health. That's when he realized he was witnessing firsthand the disastrous toll a hotter planet is having on children's well-being in

Africa and Asia. Low-lying countries like Bangladesh are experiencing widespread flooding due to the increasing frequency of torrential rains and sea-level rise, and India is subject to famines because of the droughts caused by changes in the monsoon cycle.

"I was really shocked when I began to think through the implications—it was like getting a bad diagnosis," he told me. "But we don't have much time because current climate developments are at the very worst end of the computer model predictions. It's going to seriously affect my children's future in ways that we cannot predict completely, and the future is very uncertain and potentially catastrophic. Every year of delay increases the costs and difficulties of effective action. We need a 21st-century international public health movement. Unless this challenge finds its way to the top of the international public health agenda, this century could be a disaster movie without a happy ending."

CHAPTER ONE

FEVER PITCH

————

Susan Fisher-Hoch had seen it all before. The British-born physician was no stranger to grave personal peril, having spent much of her career fighting microbial killers in some of the most primitive and forbidding places on Earth. She and her epidemiologist husband, Joseph McCormick, are celebrities of sorts in the world of virus hunters; their long and storied careers include more than a decade in the CDC's ultra-high-tech "hot zone" labs handling the world's most dangerous pathogens. They were routinely dispatched at a moment's notice to global trouble spots—battling outbreaks of Ebola fever and Lassa fever in remote villages in Nigeria, where they used primitive equipment to treat hundreds of terrified patients in understaffed bush hospitals, and Crimean-Congo hemorrhagic fever in South Africa and Senegal. In the early 1990s, they established a molecular biology research laboratory in Karachi, the largest in Pakistan, to study pathogens that plague developing nations and ventured to nomadic settlements in the desert to thwart outbreaks. The two doctors then moved on to Lyon, France's second-largest city and the headquarters of Interpol, to oversee the design and building of a Biosafety Level 4 lab, a research facility where scientists encased in biosafety suits reminiscent of those astronauts wear can decipher the secrets of humanity's deadliest enemies.

At the turn of the 21st century, Fisher-Hoch and McCormick landed in Brownsville, a bustling Gulf Coast port of almost 200,000 people at the southernmost tip of Texas. They had taken on the formidable task of helping to set up a new campus for the University of Texas's School of Public Health. It was meant to be a retirement of sorts for the veteran disease

detectives, then in their early 60s—a place where they weren't in danger of
succumbing to the maladies they were fighting and could live in the relative
comfort of an American city, far from having to wash up in a bucket of
filthy water after traveling all day on dusty roads in the scorching heat. In
Brownsville, they planned to pass on their hard-won insights to the next
generation of young scientists. Fisher-Hoch, who studied French at the
Sorbonne before she decided to become a doctor and still speaks with the
clipped lilt of her native England, never expected to be thrust once again
into the middle of an exotic disease outbreak as potentially deadly as anything she had encountered in the ghettos and rain forests of Indonesia—
right here in the American South, no less. Her account of the public health
threat that unfolded over the summer of 2005 is a chilling harbinger of the
plagues to come with climate change, and a lesson in how quickly such epidemics can spread.

On a sticky and hot day in late June of that year, a young woman in
Brownsville came down with what seemed like a bad case of the flu. Her
body was racked by chills, she couldn't stop vomiting, her blood pressure
was perilously low, and she was urinating blood. She was admitted to the
local hospital, where doctors pumped her full of fluids to treat dehydration
and dosed her with antibiotics, then sent her home three days later, when
she seemed better. But a week after she had shown up at the hospital, blood
tests that finally came back from the regional Border Infectious Disease
Surveillance project unmasked a surprising culprit: dengue hemorrhagic
fever, a deadly mosquito-borne viral disease once exiled to the tropics. The
Texas woman would become Patient Zero, the medical sentinel whose illness was the first ominous sign of a potentially serious disease outbreak.

Over the years, medical experts in Brownsville had seen dozens of cases
of the disease's milder cousin, dengue fever, but typically in people who had
been traveling—in Mexico mainly, or in one of a half-dozen countries in the
Caribbean and South America where the illness is now rife. But this was the
first documented case of the more serious form of dengue in a Texas resident who'd been infected in the continental United States. The deadly
invader, which can trigger massive internal bleeding by eating away blood
vessels and internal organs, had officially established a beachhead here.

Patient Zero recovered, but she was lucky: Other victims around the globe have found themselves beyond the help of modern medicine, and even if they are tethered to a spiderweb of life-support equipment, some die from severe internal bleeding and shock. Fisher-Hoch and other health authorities were alarmed at the possibility of a deadly dengue epidemic because there is no vaccine to protect against it or medicine to cure it. More than a century before, in the 1880s, when Brownsville was mainly an Army garrison guarding the southern border, the city—with its tropical climate of mild winters and hot, humid summers—had been gripped by an epidemic of another mosquito-borne scourge: yellow fever, a cousin of dengue and a hemorrhagic viral disease that kills three-quarters of its victims by causing massive organ failure. The grave worry was that "if dengue moves into a population that hasn't experienced it and has no natural immunity, we were going to see more severe disease," Fisher-Hoch told me during a recent interview. "The way we live hermetically sealed in our houses—with screens on the windows and air-conditioning—protects us enormously. But what about those who don't live well—the people in trailer parks sitting on their porches surrounded by mosquitoes? The potential picture was not pretty."

American health officials joined their counterparts just across the Rio Grande in Matamoros, a Mexican city with a population of nearly 750,000, to do the gumshoe sleuthing of epidemiology. They dispatched teams to carry out a blood-sampling survey to uncover the overall extent of dengue infections. They knocked on doors throughout Brownsville, in Matamoros, and in the colonias, the squalid shantytowns that line both sides of the border. There, residents live in cramped quarters with poor sanitation. The old tires, rusty buckets, and plastic containers that litter the encampments collect stagnant water, making them ideal breeding grounds for mosquito larvae. "They don't have running water or paved roads," Fisher-Hoch said. "People store water outside, creating an environment in which the mosquito is going to flourish."

Health officials discovered that nearly 1,300 people had been infected with dengue fever by the time the outbreak peaked in December, including 20 who were stricken with dengue hemorrhagic fever, though there were no apparent fatalities. The results of a random blood survey were even more startling: 76 percent of the residents of Matamoros carried dengue antibodies, which means they had been infected with the virus at some point in their lives, even if they never showed signs of infection. Worse yet, they found evidence of past dengue infection—despite the better protection afforded by window screens and air-conditioning—in nearly 40 percent of those tested in Brownsville, the highest prevalence of anti-dengue antibodies recorded in the continental United States in the previous 50 years. And of the 24 Brownsville residents who had never traveled outside the United States, 6 tested positive for dengue—which meant the illness was now firmly entrenched within US borders.

Even in communities where infections are endemic, dengue's threat is not always obvious, mainly because its symptoms are so varied. In up to 80 percent of cases, people present with what looks like garden-variety seasonal flu—fever, aches, fatigue. Those with more severe cases experience excruciating muscle and joint pain, high fever, rash, and gastritis. Prior infection with dengue—even without apparent sickness—puts them at risk for contracting dengue hemorrhagic fever, which can be life threatening. "About 1 percent or less get clinically ill," said Fisher-Hoch. "If you've got 1 case, that means there's 100 out there that you don't know about. That's why the prevalence is so high."

Since then, dengue fever has stricken more than 6,000 people along the Texas–Mexico border, including several cases of the more deadly hemorrhagic version, and more than 10,000 cases have been reported around the United States in the past decade, according to the CDC, which also found that hospitalizations for dengue fever have tripled during the same period. Most Americans aren't even dimly aware of dengue. But rising temperatures create an incubator for transmission because mosquitoes that spread dengue thrive in warming habitats. Epidemic outbreaks throughout Central and South America—in Brazil, Mexico, Honduras, Paraguay, Costa Rica, Bolivia, and Cuba—now affect nearly 1,000,000 people annually. Thousands of cases have been reported with increasing frequency in Hawaii, Puerto Rico, and Florida, and dengue infection has been confirmed in almost every state, even as far north as Maine, Minnesota, and Washington.

Dengue was long thought to have been eradicated in the United States. Just a few generations ago, when Franklin Roosevelt was in the White House, bug- and animal-borne ills like malaria, yellow fever, Rocky Mountain spotted fever, typhus, and African sleeping sickness were prevalent in the United States. Thousands of people died in yellow fever epidemics in tropical cities like New Orleans, and the virus appeared as far north as Philadelphia, New York, and Washington, DC, which became a ghost town during the summer, when residents fled seasonal outbreaks. Malaria plagued impoverished rural areas in the Deep South until the late 1940s, when the federal government launched an eradication program that included draining swamps and carpet bombing mosquito breeding grounds with DDT, a highly potent insecticide. "People came in with lots of gold braid on their shoulders and made all sorts of promises, but then we all became aware that DDT was not such a good thing," said Fisher-Hoch, alluding to the compound's later-discovered link to cancer and threats to wildlife, which resulted in the 1972 ban on the pesticide's use in the United States. Since then, however, dengue fever has crept back into a sizeable portion of the Western Hemisphere because, said Fisher-Hoch, "the mosquito reestablished itself, and dengue follows the vector."

Over the last half-century, as the planet has experienced a warming trend, dengue has spread into more temperate areas. In that time its incidence has spiked 30-fold, according to the World Health Organization, and it now causes an estimated 100 million infections annually in more than 100 countries, especially in densely populated and developing megacities in the tropical belt, where a high percentage of the population lives in urban shantytowns. The *Aedes aegypti* mosquito that transmits dengue is a sociable urbanite that feasts mainly on humans and likes to nest in moist places where it can reproduce, such as damp cellars, hidden pools of water at the bottoms of wells, storm drains, flowerpots, and garden fountains. Because of the speed of its spread and the overwhelming burdens of illness and death it causes, the WHO considers dengue the world's most serious insect-transmitted viral disease. But many doctors are unfamiliar with the symptoms and fail to make an accurate diagnosis. As a consequence, the CDC believes many cases are never counted, making these figures underestimates of its prevalence.

A range of factors influences the spread of dengue, but global warming is a particularly important driver because hotter weather triggers a chain reaction of changes in delicately calibrated ecosystems. For starters, the mosquitoes that transmit dengue fever are exquisitely sensitive to temperature changes: Frost kills both adults and larvae, so warmer winters enable mosquitoes to survive in places that once were too cold. Since the middle of the 20th century, two species of mosquitoes capable of transmitting dengue fever, *Ae. aegypti* and *Aedes albopictus* (an insect that's also called the Asian tiger mosquito), have expanded their range into newly temperate areas in at least 28 states in the United States, even as far north as New York and New Hampshire, making the 173 million Americans who live in these areas vulnerable, according to a July 2009 report by the Natural Resources Defense Council (NRDC). "Milder winters, hotter, wetter summers, and even droughts can bring this insect-borne threat closer to home," said a senior scientist at the NRDC, Kim Knowlton. "Usually relegated to tropical and exotic locales, dengue fever has rarely been an issue domestically. But a changing climate may allow dengue-spreading mosquitoes to flourish in nearly half of the United States."

Warmer weather extends the disease-transmission season because the breeding cycles of mosquitoes shorten, which means the bugs can reproduce multiple times instead of once or twice—resulting in even more mosquitoes to spread disease. Heat also speeds up the incubation of the dengue virus inside the bugs, so it becomes infective much faster, giving it a longer time to sicken someone during its three- to four-week life span. Female mosquitoes bite more frequently when the thermostat rises, boosting their capacity to transmit the virus. Moreover, hotter weather amplifies the impacts of hurricanes and storms, causing changes in rainfall patterns and flooding, which in turn expand the habitats hospitable to mosquitoes. Storms and floods disrupt shelter and water, sewage, and sanitation systems—the very things that have insulated Americans from contagions.

The fact that dengue has penetrated beyond border hot spots like the Brownsville–Matamoros area illuminates the extent of the invasion of the highly aggressive Asian tiger mosquito, which can transmit not only dengue, but also yellow fever and several types of encephalitis viruses. First found in the United States in the 1980s—most likely having stowed away in recycled tires imported from Asia—the mosquito has now been found as far north as New England, according to University of Florida research.

The unintended consequence of the public health triumph against mosquitoes in the last century was that prevention and vector-control strategies

were largely abandoned, and the surveillance infrastructure for these pathogens steadily eroded. But in hotter tropical climes, especially in overcrowded
African, Asian, and South American megalopolises lacking adequate sanitation, the tiny insects continue to act as airborne angels of death, delivering
payloads of lethal pathogens that sicken 500,000,000 people and claim at
least 1,000,000 lives annually. The number of people vulnerable to these
afflictions, both here and in the developing world, is expected to soar into
the billions as the planet heats up and mosquitoes move to higher latitudes.

And it's not just mosquitoes that are expanding their ranges. Ticks, mice,
and other carriers of potentially deadly microbial hitchhikers are surviving
milder winters and moving into newly warm regions. The increased weather
variability that accompanies rising temperatures may be responsible, at least
in part, for the spike in such tick-borne illnesses as babesiosis, a potentially
lethal flu-like infection called HGA (human granulocytic anaplasmosis),
and Lyme disease, the number one vector-borne disease in the United
States, whose incidence has doubled since 1991 and now affects nearly
325,000 annually—more than AIDS, West Nile, and avian flu combined.
The hardy black-legged ticks that spread the Lyme-causing bacteria have
taken up residence along the Canadian border, while Lone Star ticks, southern natives that transmit ills like tularemia and HGA, have been spotted as
far north as Maine. Said Louis Magnarelli, an entomologist and director of
the Connecticut Agricultural Experiment Station in New Haven, which is
near where Lyme disease first surfaced in the 1970s, "Many decades of a
warming trend with less snowfall would lengthen the infective season for
ticks and favor the survival of white-footed mice, whose numbers now drop
dramatically during the harsh winters."

Hot, Flat, and Wet

These climate-driven changes in the ecosystems of disease vectors don't
occur in a vacuum. Other trends have accelerated them, spreading contagions into populations with no natural immunity. The explosion of international travel on a hotter, wetter planet—more than 60 million Americans
travel abroad every year, and an equal number visit the United States—has
created the perfect conditions for the increased transmission of lethal

pathogens from the tropics to industrialized nations. Hitchhiking parasites and infected individuals carrying microbes that can be passed on by mosquitoes can now go anywhere in the world in less than 24 hours and deliver reservoirs of malaria, dengue, or chikungunya fever, a particularly nasty infection that causes arthritis-like joint pain, to newly temperate regions. In a hot, flat, and crowded world, to use the phrase coined by Thomas Friedman, microscopic invaders are just a plane ride away. "When you have this combination of tremendous global transport of people and goods, changing ecosystems, and changing climate," said Dr. John Balbus of the NIEHS, "it increases emerging diseases."

These two factors—global movement and changing global weather—are what enabled the West Nile virus to become entrenched in North America. The deadly pathogen first emerged in the West Nile district of Uganda in 1937. It is fairly common in Africa, West Asia, and the Middle East, and occasionally flares up into epidemics, such as those in Israel in the 1950s and in Romania in 1996. Each of these outbreaks occurred after an unusually dry, hot spell, according to research by Israeli scientists, creating the perfect incubator for *Culex pupiens*, a common house mosquito that transmits West Nile. Most people infected with the virus don't become ill, or they may experience mild flu-like symptoms such as a headache, sore throat, and fatigue. But in the elderly, young children, or people with compromised immune systems, West Nile can be a killer, attacking the nervous system and causing encephalitis—a severe swelling of the brain—that can lead to tremors, seizures, paralysis, coma, or even death, as well as other complications such as pneumonia. Some victims are permanently disabled with severe musculoskeletal or neurological damage.

In 1999, the virus arrived in New York, most likely lurking in a mosquito stowed away on a plane or in the bloodstream of someone already infected. Epidemiologists think it may well have been an Israeli from Tel Aviv—which was in the midst of an outbreak that had sickened hundreds—who was visiting relatives in Whitestone, a leafy middle-class suburb perched on the Long Island Sound in northern Queens. According to that scenario, the human carrier was bitten by a local mosquito, which then bit someone else and passed on the virus. "People's behavior was conducive for transmission," recalled Duane Gubler, who directed the CDC's Division of Vector-Borne

Infectious Diseases at the time of the outbreak. "They sit out on the back porch in the evening, smoking their cigars. It's a perfect time for transmission by mosquitoes, which like to bite at dusk."

Once again, the summer was unusually hot and dry in New York—the hottest July on record at that time—and the high temperatures helped West Nile gain a foothold. "By mid-August, there were 10 days over 100 degrees that were followed by heavy rains and flooding, which made the mosquito populations explode," recalled Tracey McNamara, a veterinarian who was head of the pathology department for the Wildlife Conservation Society at the Bronx Zoo during the initial West Nile epidemic.

In early August, McNamara sensed something was gravely amiss: She noticed dead crows on the ground. As the weeks passed, more dead carcasses littered the zoo's expansive landscape. American crows are hardy survivors, and this sudden, inexplicable die-off was virtually unprecedented. Puzzled, McNamara sent the dead birds for analysis at the New York State Department of Environmental Conservation (DEC), which is located in Albany, the state capital, about 150 miles north of New York City. She later learned that throughout the summer various state and local agencies had sent hundreds of dead crows to the department's laboratory, but they had sat in the facility's refrigerators. "I called the lab in Albany every day for two weeks complaining that I wasn't getting any results," McNamara told me when we talked over lunch near the campus of the Western University of Health Sciences in Pomona, California, where she is now on the veterinary school faculty. "But they only did a cursory exam. Then I found out the birds had been dying since June and they had 400 dead crows sitting there."

Soon, press reports started appearing, saying there was some kind of bird plague spreading through northern Queens, and that local sanitation workers and highway cleaning crews were finding hundreds of dead robins and crows on sidewalks and city streets and in the parks. By Labor Day, whatever was killing the crows was infecting the zoo population: A flamingo and a cormorant had died, and at least a dozen more zoo birds were ill. Frustrated, McNamara decided to perform her own autopsies on the two exotic birds. When she opened them up, what she found shocked her—something had caused the birds' hearts to swell and become inflamed, and massive brain bleeds, worse than anything she had seen in nearly 20 years, had occurred. "When I saw the hemorrhages in the flamingo, my heart sank and I hit the zoo's tissue bank library," recalled McNamara, who told me that at that moment, she came to the chilling realization that she was dealing with something that might kill her, too.

McNamara suspected that a virus spread by mosquitoes was the probable miscreant. Stagnant pools of water swarming with mosquitoes must have attracted the birds and they became infected, and in turn transmitted the virus to other birds. But it was a holiday weekend when she came up with this theory, and the person she normally sent samples to at the Albert Einstein College of Medicine, also in the Bronx, was on vacation. So McNamara donned protective gear and three sets of gloves. Hunched over a microscope enclosed in a filtered-air safety cabinet, she sliced off tiny slivers of tissue from each organ and carefully placed them on dozens of slides. Sitting alone in the deserted lab, McNamara resolved to call her lawyer after the holiday weekend to write her will.

The day after Labor Day, several more birds died. "The virus swept through the bird population at the zoo," she recalled. "Within a couple of weeks, 24 zoo birds had to be euthanized. Even asymptomatic cranes had puffing in the brain." McNamara told city health officials about her suspicions and shipped them tissue samples, then called the CDC. "I couldn't get the CDC to listen," she said. "Terrified and beside myself, I literally woke up at two in the morning and thought 'I'll call the army'"—the US Army's top secret biological warfare laboratory at Fort Detrick, Maryland, where researchers search for ways to thwart deadly infections. McNamara had worked with scientists there during the 1989 Reston, Virginia, Ebola outbreak in a group of monkeys imported from the Philippines, which was chronicled in the best-selling *The Hot Zone*. "The army went ballistic and couldn't believe I wasn't getting any cooperation," she said. "They told me to continue doing necropsies and sent me biosafety equipment."

Meanwhile, a family physician at Flushing Hospital Medical Center in Queens had a cluster of five elderly patients in the intensive care unit who were deathly ill from a mysterious neurological disease that was causing encephalitis. They seemed confused, complained of headaches and severe gastrointestinal pain, and experienced muscle weakness and fevers as high as 103°F. Three other patients with similar but less severe symptoms showed up shortly afterward. Ultimately, 62 people were hospitalized, 7 of whom died. Routine diagnostic tests hadn't been able to identify the culprit. Encephalitis is a relatively rare disorder. Even big-city hospitals get only a few cases a year—the entire city of New York normally reported about 10 cases annually—and it was highly unusual for a community hospital to see several patients at once. The doctor notified the New York City Department of Health about her concerns. But West Nile virus had never been seen before in the Western Hemisphere, so public health officials had no reason to suspect it when the virus claimed its first victims.

After the patients died, their brain tissue samples were shipped to the laboratory of the vector-borne disease branch of the CDC in Fort Collins, Colorado. There, scientists made a tentative diagnosis: Saint Louis encephalitis (SLE), which, as the name suggests, is caused by a virus that lives in birds and mosquitoes in the Midwest and South. Mayor Rudy Giuliani immediately ordered a concerted aerial-spraying campaign to eliminate the mosquitoes that were believed to be spreading this new disease. That September, helicopters could be heard all over the city and in many suburbs, saturating the landscape with pesticides used to eradicate the disease-carrying insects.

But there were puzzle pieces that still didn't fit. "The SLE virus never kills birds, and here Tracey comes up with something that was killing her birds," Duane Gubler of the CDC told me. Perplexed, scientists went back to the bench to decode the virus's DNA and made an astonishing discovery: It was West Nile. "That's when I said, 'Holy shit,'" Gubler recalled. "West Nile was the last thing we were expecting. The infectious disease world was pretty complacent in those days, but that was a wake-up call. We quickly learned that you don't just focus on the bugs that you know about in the region. In today's world of globalization, it can be anything."

But more than two weeks had passed since McNamara had called the CDC, a lapse that illuminated the serious deficiencies in our disease-surveillance network. At first, public health officials didn't link the bird and human outbreaks. McNamara's heroic efforts to get their attention were "initially dismissed by many involved in the investigation," according to a critical 2000 report from Congress's General Accounting Office, a failure that led to a damaging delay in correctly identifying the pathogen. "Rapid and accurate diagnosis of outbreaks is essential," according to the GAO, so that prevention and control measures can be speedily implemented, victims can be given the most effective treatment, and health care and lab workers who are exposed to these pathogens can be adequately protected. McNamara remains convinced that lives were lost because there was little communication between governmental agencies, and that if adequate tests had been done when the virus started killing birds, mosquito eradication efforts would have commenced sooner. "There were 400 dead crows that had been sitting in DEC's refrigerators in Albany since June," she recalled with no small measure of outrage, still shaken more than a decade later by what she viewed as bureaucratic incompetence bordering on criminal negligence. "Birds, dogs, cats, and other animals are urban disease sentinels. In this case, taxpayers became the sentinels. To this day, I don't think lessons have been learned."

When another hot, dry summer hit in 2002, the disease spread across the United States and into Canada. West Nile is now firmly entrenched because warmer temperatures have allowed mosquitoes to survive during the milder winters. Since the initial 1999 outbreak, at least 35,000 people in North America have been infected, and there have been more than 1,300 fatalities. In 2012, during another hot, dry summer, West Nile emerged again. Every state in the continental United States reported infections, although Texas was the hardest hit, with more than 4,500 people sickened, 183 of whom died. But the virus is not the relatively innocuous disease that public health officials would have us believe: It may kill up to 15 percent of its victims. "It's in breast milk, in the blood supply, and in organ transplants," said McNamara. "They say that West Nile only hits the young and the elderly. But there are people in their 40s who ended up in a long-term care facility."

Could malaria and other exotic diseases that rarely occur in the industrialized nations of North America and Europe follow a similar path? Dr. James Diaz, master's program director in environmental and occupational health at Louisiana State University School of Public Health in New Orleans, thinks so. He points out that mosquito carriers for dengue, yellow fever, and chikungunya virus are already present in the United States. "They're simply awaiting an opportunity to transmit these diseases locally between arriving infected airline travelers and nearby residents," he told me during a recent interview at his office in downtown New Orleans. "We think we're protected by the Atlantic and the Pacific Oceans, but the reality is that we're one little border stop or an eight-hour plane ride away from someone who has just flown into London from Africa."

Many of these diseases can become established before we're even aware of them. "West Nile may have been here long before they recognized it, and it could have been transmitted among the bird population before it reached that threshold where it started infecting humans," said Kerry Clark, an expert in vector-borne diseases at the University of North Florida in Jacksonville. "We'll probably have some small focal outbreak from an imported human case that goes unrecognized for a while and—like a smoldering fire that eventually blows up—the next thing you know, malaria is endemic in the United States."

RETURN OF THE MEGADEATH

The wild swings in weather that are expected to become commonplace on a warmer planet—more frequent and severe droughts followed by drenching rains—also change ecosystems in a way that awakens and expedites the transmission of once-dormant pathogens. The emergence of hantavirus pulmonary syndrome, a deadly malady spread by rodents, is a good example. In the spring of 1993, within the remote Four Corners region of the Southwest, there was an outbreak of a mysterious respiratory illness that was killing young and otherwise healthy people. Doctors were stumped. Patients would show up at hospitals with mild flu-like symptoms, but the illness would swiftly turn lethal: Their lungs would rapidly fill with fluids, rendering them unable to breathe. The death rate hovered at around 40 percent, even for those who made it to the hospital as soon as they fell ill. Within six months of the initial outbreak, more than 60 cases were reported, of which 25 were fatal. One patient who was sitting up in bed in the morning, talking and eating breakfast, was gasping for air by the afternoon and dead before nightfall, essentially having drowned in his own fluids.

After some intensive medical detective work, health authorities from the CDC and local agencies in New Mexico and Arizona determined the disease was caused by a previously unknown type of hantavirus that is carried primarily by the deer mouse. Hantaviruses are common in Asia—the group is named for the Hantaan River in Korea, where the illness first emerged—and seriously sickened about 200,000 people each year, mainly in China around harvest time, when rat burrows in the fields get stirred up and release dust containing aerosolized rodent urine. Baffled scientists tried to unravel why the virus would suddenly emerge and cause the cluster of cases in the sparsely populated Four Corners, where the boundaries of Utah, Arizona, New Mexico, and Colorado meet.

Initially, they knew that the virus was carried by rodents—but which species? They left traps around the homes of Four Corners residents and caught 1,900 rodents in all, which they then dissected. Tissue and blood samples were packed in dry ice and shipped to the CDC for analysis, which revealed that the common deer mouse was the carrier. But these cute little rodents have inhabited the Southwest for centuries. Why would they cause an outbreak now? Then CDC officials discovered there were 10 times more deer mice in the Four Corners in May of 1993 than there had been in the previous year, which led them to conclude that a recent dramatic shift in the weather likely created the conditions that triggered the fatal outbreak.

The region had been crippled by severe drought for six years, sharply culling the population of predators and leaving rats and mice fighting each other over what little food was available (and spreading the virus among themselves through their urine and saliva). But the change to an El Niño climate cycle, which occurs when the ocean's surface temperature warms, had triggered massive rains and snowfall in the late winter and spring of 1993. This produced a bounty of the seeds, nuts, berries, and insects that the rodents feed on, spawning legions of mice. More mice boosted the chances of human exposure to infected rodents. Health officials suspect a similar climate cycle—a mild winter that allowed adult rodents to survive, followed by a wet summer—played a role in the 2012 hantavirus outbreak in Yosemite National Park, in which six people were infected, two of whom died, and literally thousands of campers from all over the world were exposed.

Intriguingly, this type of weather pattern may be what led to the fall of the once mighty Aztec Empire in the early 16th century—and not, as is commonly held, the invasion of European colonialists, who brought with them diseases like mumps, measles, and smallpox to which the native populations lacked immunity. When Hernando Cortés and his army began conquering Mexico in 1519, there were roughly 25 million people living in what is now Mexico. A hundred years later, after a series of epidemics decimated the population, perhaps as few as 1.2 million natives survived. Records confirm there was a smallpox epidemic in 1519 and 1520, immediately after the Europeans arrived, killing between 5 million and 8 million people. But it was two cataclysmic epidemics in 1545 and 1576—26 and 57 years after the Spanish invasion—that swept through the Mexican highlands and claimed as many as 17 million lives. Mexican epidemiologist Dr. Rodolfo Acuña-Soto called this "megadeath" toll "one of the worst demographic catastrophes in human history, approaching even the Black Death of the bubonic plague, which killed approximately 25 million in Western Europe." However, Acuña-Soto was skeptical that smallpox was the sole culprit behind the Aztecs' annihilation.

I caught up with Acuña-Soto, a physician and Harvard-trained infectious disease specialist, at his office at the department of microbiology and parasitology in the medical school of the National Autonomous University of Mexico in Mexico City. Not long after he returned to his native country after completing his PhD in tropical medicine at Harvard, he became obsessed with solving the mystery of what had actually killed the Aztecs. It just didn't make sense that a deadly outbreak of European origin could occur so long after the Spanish had arrived, he said, because the natives who had survived previous plagues would have passed on their immunities.

Fortuitously, the Spanish conquerors kept detailed records that had been meticulously archived for centuries in Mexico and Madrid. To find some answers, Acuña-Soto spent more than 12 years poring over the documents of the 16th-century Spanish priests who had worked with the Aztecs to preserve a record of their history, language, and culture. These texts also tracked key natural events—storms, droughts, frosts, and illnesses. In particular, they detailed the plagues of cocoliztli ("pest," in the Aztecs' native Nahuatl), which appeared to have been far more virulent than smallpox. "Nobody had the health or strength to help the diseased or bury the dead," one Franciscan friar wrote in 1577. "In the cities and large towns, big ditches were dug, and from morning to sunset the priests did nothing else but carry the dead bodies and throw them into the ditches."

Acuña-Soto also gained access to exhaustive diaries kept by Francisco Hernández, personal physician to King Philip II and the equivalent of surgeon general of the territory now called New Spain. After witnessing the second catastrophic epidemic in 1576, Hernández interviewed hundreds of Indians and performed autopsies on many of the victims. He described a highly contagious and lethal scourge that killed within a few days, causing raging fevers, jaundice, tremors, dysentery, abdominal and chest pains, thirst, delirium, and seizures. "Blood flowed from the ears," Hernández observed, "and in many cases blood truly gushed from the nose."

"These symptoms didn't sound like smallpox or any other known European disease that was in Mexico during the 16th century," Acuña-Soto said. "This sounded like a hemorrhagic fever. So if the Spanish didn't bring about the fever, what did?" In his research, Acuña-Soto noticed a pattern: The plague appeared during a period of wet weather and heavy rainfall that followed years of severe drought. To confirm his observations, Acuña-Soto worked with a Mexican–American team of dendrochronologists—scientists who study tree rings to date changes in climate—and compared the 16th-century historical accounts with tree-ring records from a forest of 450-year-old Douglas firs in a remote region of central Mexico. Thick tree rings indicate wet years, when plants flourished, and thin tree rings reflect lean, dry years. These rings indicated that the most severe and sustained drought in North America in the previous 500 years occurred in the mid-16th century and extended from the Pacific to Atlantic coasts, encompassing nearly the entire continent. But there were heavy downpours in the periods around 1545 and 1576—the years of the cocoliztli outbreaks. "The smoking gun was the tree-ring data," Acuña-Soto said.

Acuña-Soto is now convinced that the death knell for the Aztecs was an indigenous hemorrhagic fever virus spread by rats, not the European diseases that accompanied the Spanish conquest. Much like the transmission pattern of the hantavirus, the rat population was depleted during the drought, when food was scarce. Once the rains returned, food and water were suddenly plentiful, so the number of infected rats exploded and the deadly scourge spread to humans. In our own future, as weather swings become more severe and erratic and the Southwest bakes during increasingly severe and prolonged droughts, epidemics like *cocoliztli* will doubtless return. "Sooner or later, a new virus will emerge from animals, from the desert, for which we don't have any vaccine [or] drugs," said Acuña-Soto. "That's guaranteed. That's the big fear of science. The only thing we don't know is when."

GOING TO WAR

Brownsville is now at war with dengue fever. Mosquito traps have been set up all over town, and vector-control teams are routinely dispatched to mosquito hot spots to spray insecticides and apply larvicides, especially after heavy rainfalls. "We're not going to bring DDT back, but if you understand what the risks are, you can take the proper precautions," said Fisher-Hoch. The local health department is part of the Binational Health Council, a consortium of fifteen health agencies along the US–Mexico border that coordinates surveillance, education, and containment efforts. Neighborhoods on both sides of the border have been blanketed with bilingual pamphlets providing mosquito prevention tips, educational seminars are conducted at local schools, and training sessions are held at the region's medical centers to teach doctors how to detect and treat dengue. "Physicians coming into the area aren't familiar with dengue, and tropical diseases need to be part of their diagnostic workups," said Brownsville's public health director, Arturo Rodriguez, who perhaps not coincidentally was one of Joseph McCormick's first graduate students at the newly opened local campus of the University of Texas's School of Public Health. "We also meet regularly with our counterparts in Mexico and have jointly made dengue one of our top priorities in health."

These collaborative efforts are part of a larger campaign by local health departments and the CDC to improve their ability to prevent, detect, and respond to disease outbreaks in an era when "infectious pathogens are rising from apparent defeat to wage a new war on mankind," according to Duane Gubler. The CDC is harnessing the latest technology to enhance disease surveillance at key entry points on our borders to prevent lethal pathogens like dengue from taking root and to swiftly identify disease-causing microbes so that interventions can be started before outbreaks spiral out of control. The agency's recently launched National Environmental Public Health Tracking Network is an online database that collects information from a variety of national, state, and municipal sources, including health departments and hospital emergency rooms, to track disease outbreaks and deliver real-time surveillance on emerging public health threats when time is of the essence and lives are at stake. They're at "the beginning of a long-term process building up our national surveillance network for these diseases, which was a huge task," said Lyle Petersen, director of the CDC's Division of Vector-Borne Infectious Diseases. "Now we're developing specialized sites around the country to home in and investigate these diseases in much greater depth."

For example, the agency has partnered with the Mexican Secretariat of Health to establish the Border Infectious Disease Surveillance project, known as BIDS, an early-warning and disease-monitoring network that pools the resources of health departments in cities on both sides of the 1,500-mile border, including San Diego, Sonora, Ciudad Juárez, El Paso, Nogales, and Las Cruces. They certainly face a challenge: It's the busiest international border in the world, with up to 400 million people legally crossing northward every year, and the region contains five of the seven poorest areas in the United States and an estimated 3,000 colonias, or unincorporated towns.

More than 400,000 Texans live in these colonias, which are veritable disease hot spots because they have little access to clean drinking water, sewer systems, sanitation, electricity, or most other modern conveniences. Started in the 1950s by predatory developers that sold worthless land with little or no existing infrastructure to mostly Hispanic immigrants working on farms in the United States, the colonias grew rapidly in the 1980s and 1990s, especially after the North American Free Trade Agreement moved thousands of jobs south of the border, bankrupting many families that had depended on relatively well-paid factory work. Many of the colonias flood frequently because of poor drainage, there is no trash removal or adequate plumbing, and homes lack air-conditioning and window screens. Mosquitoes breed in containers used to store drinking water and in old tires, broken ceramic jars, potholes in the unpaved roads—anyplace where stagnant water accumulates. These deplorable conditions allow infectious diseases such as malaria, cholera, hepatitis A, dengue, and Lyme disease to spread unchecked.

"Infectious disease incidence is higher in border regions that reflect the movement of people from Mexico and other countries in Central and Latin America," said Dr. Stephen Waterman, who helped set up BIDS in the CDC's Division of Global Migration and Quarantine Border Health Services Station in San Diego to deal with the massive movement across borders. BIDS, which played a key role in spotting dengue viruses as the cause of the 2005 Brownsville outbreak, has improved disease surveillance by educating doctors and nurses on both sides of the border about what clinical syndromes to look for and what diagnostic tests need to be done, offering quick turnaround for tissue sample testing.

All these fledgling efforts to restore the basic public health safety net are laudable. Unfortunately, severe recessionary budget cuts across the country threaten to undermine our ability to combat diseases, according to a disheartening 2009 survey by the American Public Health Association. State and local health departments have reduced or eliminated epidemiological investigation, which seriously impairs their ability to detect outbreaks and swiftly mount a response. Public health laboratories have been virtually dismantled, with more than 96 percent of all federal, state, and local government facilities having experienced deep cuts in their budgets that have prompted widespread layoffs of key personnel, delayed the purchase of crucial supplies and equipment, and eliminated or reduced services that are essential to public health. "Overall, public health laboratories are less prepared than ever to respond to a large infectious or novel disease, food or waterborne disease outbreaks," the report concluded. "The reductions in the public health infrastructure make it increasingly difficult to detect and track emerging diseases," complicating cities' efforts to prevent or stop the spread of disease.

What's worse is that there is virtually no international network that is able to contain, much less prevent or detect, the kind of large-scale global health emergencies that are more likely to occur as the thermostat rises. Former CDC staffer Duane Gubler could see these ominous trends when he retired from the agency more than a decade ago. Under the auspices of the

Duke University–National University of Singapore Graduate Medical School, he has spent the time since building the scaffolding for an international disease research and surveillance network to thwart the next global pandemic. Called the Emerging Infectious Disease Program and based in Singapore—a stable, prosperous nation that is the financial and shipping hub of Southeast Asia—the program plans to open field sites in Vietnam and Sri Lanka. The EIDP has also partnered with organizations throughout the region, like Tan Tock Seng Hospital, Singapore General Hospital, DSO National Laboratories, and the Novartis Institute for Tropical Diseases, to develop the laboratory, epidemiologic, and scientific capacity to make Singapore a surveillance center for emerging infectious diseases.

"We pour hundreds of millions of dollars into Africa, but the emphasis on Africa is mostly due to humanitarian and political pressure. In terms of the global emergence of infectious disease, it's relatively unimportant," said Gubler, now a professor at Duke–NUS Graduate Medical School. Virtually all of the recent global health scares—SARS, Hong Kong and avian flus, and Japanese encephalitis—originated in Asia, he noted. "If you look at global trends, almost all of the economic growth and all of the urban population growth in the next 20 years will occur in the countries and cities of Asia. Migrant workers come from rural areas to cities to work, then go home to plant their crops, which greatly increases the introduction of zoonotic [animal-borne] pathogens into an urban environment. Add to that the poor housing and inadequate basic services, and you have the ideal conditions for transmission of all kinds of infectious diseases. And every one of these cities—Bangkok, Manila—have modern new airports and millions of people and animals move through these airports every year, many of whom travel to the West."

The swift spread of SARS from southern China to North America is a case study in how fast diseases can explode into global epidemics in this age of melting distances. The first case of infection in late 2002 was believed to be in a pig farmer in Foshan, a city in the Guangdong province, near the Hong Kong border. About 800 people were sickened and 34 died in the initial outbreak of the disease, which is spread from person to person in droplets when someone sneezes or coughs. Two months later, in February 2003, Dr. Liu Jianlun, who had treated cases in Guangdong, arrived in Hong Kong to attend a wedding and checked into a room on the ninth floor of the Metropole Hotel. He felt slightly feverish, but well enough to do some shopping and sightseeing and to dine with his brother-in-law. The next day, he was admitted to the intensive care unit of a local hospital and died 10 days later. Two weeks after that, his brother-in-law succumbed, too.

It soon became evident that the doctor had infected 12 other hotel guests and visitors. They included a tourist from Toronto, a flight attendant from Singapore, and a Chinese American businessman who subsequently flew to Vietnam. Because the highly contagious SARS virus has a long incubation period—it can take up to 10 days for people to get sick—they all carried the disease out of Hong Kong. By the end of March, all schools in Hong Kong and Singapore were shuttered, thousands of people were quarantined in China, Vietnam, Hong Kong, and Toronto, and the WHO issued an emergency travel advisory alert, crippling tourism. By the time the crisis was finally over in July, SARS had sickened 8,000 people and claimed 774 lives in 25 countries.

As climate-driven droughts and floods trigger massive migrations to the makeshift slums surrounding Asian megacities, the displaced will carry with them microbial hitchhikers like the SARS virus that jump among animal species and humans and in the past would have been contained in remote regions. "These pathogens have the potential to cause unbelievable economic and social damage because it shuts down the economy," Gubler said. "We are not prepared. We go from one crisis to another mainly because we don't have the resources to implement the kind of public health programs we need."

CHAPTER TWO

BREATHLESS

Martha Cota awakened one morning to find her infant son, Jose Miguel, gasping for air, his lips and the skin under his fingernails blue from lack of oxygen. Terrified, she rushed him to her local hospital in Long Beach, California, where doctors stabilized the toddler and sent him home with medications to control his fever and an inhaler to help him catch his breath. This horrifying cycle went on for more than five years—Jose turning blue and barely able to breathe, Cota frantically strapping him into his car seat and racing through traffic, spending countless days and nights sitting in hospital emergency rooms. Like so many other kids in her working-class neighborhood not far from Long Beach's massive port, Jose had severe asthma, and it took doctors years to hit on a regimen that controlled his symptoms. "He could have died at any time," Cota told me one April morning, her eyes brimming with tears, when we conversed through an interpreter at a coffee shop near her office in Long Beach. "At least once a month, I'd get a call from his school threatening to send me to jail because he was absent so much."

Tall and attired in an elegant black suit, Cota, who is now in her late 40s, has a long mane of dark hair shot through with auburn highlights and large, expressive eyes. She was a social worker in her native Mexico and now works as a community educator for the Long Beach Alliance for Children with Asthma. When she first discovered that many women and children in her community were having trouble with respiratory illnesses, she became determined to find out why living downwind of the nation's largest port complex was making so many people sick.

Even at noon on this otherwise sunny day, the sky was blanketed in a hazy toxic smog from the ports of Los Angeles and Long Beach, the entry points for more than half of the goods shipped into the United States and the largest source of air pollution in California. Every day, idling cargo ships carrying 30,000 containers from Asia and the Pacific Basin are unloaded by construction cranes towering more than 200 feet high over the docks. Then the containers are dispatched from the rail yards alongside the harbor onto 1,200 diesel-powered freight trains that ferry goods up and down the coast and 35,000 semi tractor-trailers that speed along heavily congested highways to the rest of the country.

Driving through the neighborhood where Cota once lived—a poor community in the shadow of the waterfront, its rutted streets lined with 1950s-style warrens of stucco-and-concrete apartment buildings and neat but tiny clapboard cottages with barred windows—I was struck by the fact that virtually every flat surface was covered in black soot. Cars, patio furniture, windowsills, even floors and tables inside homes bore the same layer of ashen grime that Cota told me thickens as the day wears on—and residents inhale with every breath. Since the 1970s, the massive port complex has exploded to five times its original size, transforming the surrounding area into what one local physician called "an environmental nightmare."

California may have a reputation as a sun-kissed paradise with some of the world's most photographed real estate—Southern California's sparkling white beaches, the celebrated rocky coastline along Big Sur, and the epic grandeur of Yosemite and the Sierras. But the reality is that the nation's most populous state is an industrial colossus—the world's ninth-largest economy—with the worst air quality in the United States, according to annual report cards issued by the American Lung Association. It's at once the epicenter of technological innovation (Silicon Valley), an agribusiness powerhouse (roughly a quarter of the fruits and vegetables America eats are grown there), and home to the nation's largest port complex, which is a key engine of the region's economy. The Los Angeles metropolitan area, with nearly two cars per household, has the dubious distinction of having the highest vehicles per capita ratio in the world; more than 12 million

cars travel on the extensive freeway system every day. But all this commercial activity exacts a heavy toll on residents' health. The exhaust-filled
industrial corridor that stretches from the ports of Los Angeles and Long
Beach to Riverside, a city of 300,000 residents and a distribution hub that
sprouted up next to the freeways where trucks and trains transporting
goods belch dark clouds of diesel dust, is what environmentalists call the
"diesel death zone."

Cota's first encounter with the environmental cause that would come to
consume her life was at a parents' workshop held at her son's school, situated
about a mile downwind from the waterfront. There, she met a team of physicians connected with the children's clinic at the University of Southern
California (USC) who were conducting a long-term study on the link
between chronic exposure to air pollution from freeway traffic and respiratory illnesses. Called the Children's Health Study, the research project
began in 1993 and eventually involved about 5,500 children in 12 communities in Southern California. Two-thirds of them were enrolled as fourth-
graders, and all were followed until they graduated from high school. Cota
immediately signed up for the study and learned how to use a particle coun-
ter, which measures particulates and toxins in the air, to take samples of
ambient air quality surrounding her home. "Every time a truck would go
by," she recalled as we sat sipping coffee, "the meter would spike off the
charts." Sometimes it measured 7,000 micrograms of particulate matter per
cubic meter, which is more than 400 times the level federal standards consider healthy.

What the USC researchers ultimately uncovered after a decade of meticulous monitoring was eye-opening. The lung development of children who
lived near highly trafficked corridors was stunted—about 20 percent
smaller than average—which greatly impaired their functioning. These
youngsters also suffered from asthma at significantly higher rates—those
living within a quarter of a mile from a freeway had an 89 percent higher
risk of asthma than kids more than a mile away; their asthma symptoms
also were worse, and, even among those with health insurance, their odds
of ending up in the hospital emergency room choking from the bad air was
triple that of youngsters in more affluent areas. Worse yet, pregnant women
living in these high-traffic areas were more likely to give birth to premature

or low-birth-weight infants, which set up the next generation for a lifetime of disabilities and developmental deficits.

But Cota hardly needed a study to tell her this. She and two of her other children later developed severe allergies and now require inhalers and an arsenal of medications just to get through the day. "No one in my family had asthma or allergies," Cota told me. She started working for the Long Beach Alliance for Children with Asthma, she said, "so no other mother has to put up with what I had to."

The situation is even worse in California's Central Valley, where a combination of farming, industry, traffic, and topography have made the air quality so bad that 5 regional metropolitan areas rank among the nation's top 10 dirtiest cities, according to the American Lung Association; Bakersfield consistently tops the list. A long, crescent-shaped lick of fertile flatlands that stretches more than 450 miles to cover 22,000 square miles (slightly smaller than West Virginia), the Central Valley is nestled between the coastal mountains on the west and the Sierra Nevada range to the east. They call it "the other California," with patches of farmland extending as far as the eye can see and hardscrabble towns originally settled by Dust Bowl refugees now covered with strip malls and fast-food outlets. Bakersfield, the birthplace of the renegade rockabilly honky-tonk of Merle Haggard and Buck Owens, is one of them. But this high desert is home to 6.5 million people—more than the populations of two-thirds of the states in the nation. There are almost a million people in the Fresno metropolitan area alone, which has been transformed in recent years from a sleepy farm town dotted with grape vineyards and a stopover for Yosemite visitors into the region's thriving cultural epicenter, crisscrossed by three major freeways.

But the growing urban sprawl, coupled with industrial growth, has yielded increasingly tainted air. Dairy farm waste, soil blown from farmlands, pesticides, industrial emissions, vehicle exhaust, and dust particles kicked up by cars and the big rigs ferrying produce down Interstate 5 and Highway 99, the valley's two main arteries, have made this one of the smoggiest places on Earth. The surrounding mountains trap these pollutants, and the stagnant air envelops the region in a perpetual cloud of haze. Throughout the several days I spent in the area, I constantly felt like I was fighting off a cold. Even with the windows tightly shut, I'd awaken with sneezing fits as my immune system went into high alert and tried to expel the foreign particles invading my respiratory tract. As I sped down the dusty freeways along seemingly endless miles of farmland, the monotony relieved only by the truck stops clustered around the off-ramps, the dirty air irritated my throat and stung my eyes, leaving them perpetually reddened and swollen. On hot days, locals told me, the toxic smog fills hospital emergency rooms and doctors' offices with children who can't breathe, and schools in Fresno fly color-coded flags to alert students to the air quality—green means it's okay to be outside, while red is a warning to stay indoors. On average, nearly four valley residents die prematurely every day because of the pollution, and experts predict that within the next few years, as temperatures continue to rise and population growth raises smog levels, one of every four children will have asthma.

"In the past 10 years, I've had to put more kids on steroids than ever before, which terrifies me," said Kevin Hamilton, a respiratory therapist and administrator for Clinica Sierra Vista, a string of medical centers throughout the Central Valley that cares for about 50,000 low-income youngsters every year. "It can be overwhelming with patients in every room on inhalers and nebulizers"—the heavy artillery of an asthmatic's medicinal arsenal. A burly, bearded man in his late 50s with a long ponytail snaking down his back, Hamilton is an unapologetic activist who wears his '60s-style politics like a badge of honor. I sought him out because he's been on the front lines in the fight to clean up the air in the Central Valley for more than a decade and was one of the architects behind the Medical Alliance for Healthy Air. In 2001, the group, part of a coalition that included the Sierra Club and two environmental justice organizations, sued the federal government because many of the region's worst polluters, including big real estate developers and corporate farmers, were exempt from portions of the 1970 Clean Air Act. In a landmark settlement, the Environmental Protection Agency agreed to start requiring air-quality permits from farms in the valley, but these mandates were later gutted by the George W. Bush White House. "These kids never get well and now we have a generation who are permanently damaged by their constant exposure to pollution," Hamilton sighed in frustration when we met in his cramped office at the clinic. "The effects are not transient. If they get pneumonia, they're more prone to end up in the hospital. They can't participate in athletics and when they get older, they're more likely to have heart disease because their bodies can't generate enough oxygen. They get sicker faster, and die younger."

Air pollution takes a toll on the region in other, less obvious ways, such as the loss of human capital, which ripples throughout the local economy. Educated people—doctors, nurses, lawyers, skilled tradespeople—are abandoning the area in significant numbers because they're worried about their families' health: It can take months to get an appointment with a neurosurgeon, cardiologist, or oncologist, and patients often have to travel hundreds of miles to coastal cities to see specialists. The brain drain of high-earning professionals reverberates with not just the loss of their skills and of the immeasurable social capital of a more sophisticated, educated population, but also in the spending that contributes to the local economy. They buy bigger houses, more luxury items, and more expensive cars, and they pay more in taxes, which support local schools and municipal services. During one five-year period, from 1995 to 2000, the southern half of the valley region lost 3,000 people with bachelor's degrees while adding 13,000 adults without high school diplomas, according to the Public Policy Institute of California. Only 12 percent of residents are college graduates, compared to 24 percent statewide; the unemployment rate in many of its 18 counties is 5 to 6 percent higher than the statewide average. Many fear the region is rapidly becoming a feudal enclave populated primarily by impoverished farmhands and agribusiness land barons who live in palatial Spanish-style estates high on the bluffs overlooking northern Fresno.

"What we're seeing now is probably just the beginning of the effects we'll experience from bad air," Dr. Jose Joseph, a pulmonologist and asthma specialist at the Fresno campus of the University of California, San Francisco, told me rather ruefully one steamy October afternoon when I visited his tidy office at the university's medical center. "In the years to come, we're going to have major increases in all types of chronic illnesses," he continued, ticking them off on his fingers: "in respiratory illnesses, in heart disease, in increases in heart attacks and strokes because air pollution increases blood clotting, and in its effects on developing fetuses—there is so much fallout from air pollution. In looking at the magnitude of the problem, we really have to do better than this."

FUTURE SHOCK

For Californians who live, work, and breathe in the state's industrial zones, the future is already here. The disastrous health effects they experience from pollution are a preview of what will happen as the planet gets hotter, carbon levels climb, and air quality worsens. To cite one example, the heat-trapping carbon dioxide that's emitted from tailpipes and factories collects over cities, creating CO_2 "domes" that shroud the urban cores in toxic clouds of pollutants. Research on air quality in New York City, Phoenix, and Baltimore shows that ambient CO_2 parts per million (ppm) levels can spike into the 400s, 500s, and 600s—right now, the global average is 393 ppm— which climate modelers predict will become the norm in 20 to 30 years. As temperatures rise and more pollutants are dumped into the atmosphere, the plume of that toxic cloud will spread like ink on a blotter, covering more land under a suffocating carbon canopy. A 2010 Stanford University study found that these domes act like pressure cookers, exacerbating pollution's harmful health effects, and may already be responsible for up to 1,000 excess deaths across the country, the equivalent of two jumbo jet crashes every single year.

Two of the chief culprits behind asthma and allergies—air pollution and smog—intensify as temperatures rise. Ozone smog, which is a mixture of all the pollutants and particles in the air, is created when sunlight cooks pollutants in the atmosphere. When the air heats up, more ozone is produced, and more ozone in turn traps more heat, exacerbating the urban heat island effect and creating a vicious cycle. Higher levels of ozone smog, toxic to the lining of the lungs, will also boost the incidence of respiratory diseases. A 2009 study done by a team of European scientists looked at hospital admission data from 12 major cities such as Dublin, London, Barcelona, Athens, and Rome from at least a three-year period. They found that for every 1°C (about 2°F) temperature increase, hospitalizations from respiratory- and asthma-related illnesses rose by 4.5 percent.

Chronic exposure to elevated levels of ozone has a serious cumulative effect. Ozone in the upper atmosphere normally forms a protective layer that shields us from the sun's ultraviolet radiation. But ground-level ozone—the chemical combustion product of factory and vehicle emissions heated by sunlight—can be toxic. Sunbelt cities like Los Angeles; Riverside, California; and Houston, with their seemingly endless sunny days, gridlocked urban sprawl, and heat-trapping stagnant air masses, contain the highest average concentrations of ozone, according to a 2009 study by University of California, Berkeley, scientists. People living in these regions, and in California's Central Valley, have a 25 to 30 percent greater annual risk of dying from respiratory diseases like pneumonia and chronic obstructive pulmonary disease than do residents who enjoy cleaner air in places like San Francisco and Seattle, where fog, rain, and cooler temperatures keep ozone levels in check.

But higher ozone concentrations are just one aspect of the problem. Rising temperatures will make bad air even more dangerous, cooking up a witches' brew of pollutants that will sear the delicate tissue lining the lungs and aggravate an astonishing array of other health problems ranging from heart disease and lung cancer to dementia. The dirty particles accelerate the thickening of arteries, which increases the chances of heart attack and stroke, and hastens a decline in cognitive abilities because less oxygen-rich blood is being pumped to the brain. One 2012 study that followed nearly 20,000 women nationwide revealed that exposure to this type of pollutant greatly speeds memory impairment and reduces concentration. And women who experienced higher levels of exposure to tainted air for longer periods of time had "significantly" sharper declines in mental acuity, the equivalent of an extra two years of aging. "The same chemical reaction that makes more ozone and goes faster when temperatures are higher also produces chemical compounds that make particles, or particulate matter, in the air," said Anthony Wexler, director of the Air Quality Research Center at the University of California in Davis.

The big particulate matter (PM)—PM_{10}—is about 10 microns in size, similar to the thickness of a strand of hair. Typically, these particles are found in windblown or construction dust and are emitted by woodstoves, fireplaces, trash incinerators, and wildfires. PM_{10}s tend to make up that thick blanket of haze that envelops urban areas, and, when inhaled, they stick to the insides of the lungs' small branches that transport oxygen to the gas-exchanging tiny sacs called alveoli. The alveoli are surrounded by thick networks of blood vessels. This is where the crucial switch is made and our bodies perform their miraculous life-sustaining alchemy: Carbon is removed from the blood to be expelled from the lungs and replaced by fresh oxygen, which is then pumped to the heart for circulation. But the pollutants cause the lungs to make mucus, trapping these particles and creating a persistent cough.

Even finer particles, called $PM_{2.5}$s, are 2.5 microns or less in size—about 30 times smaller than the diameter of a human hair—and invisible to the naked eye. These tiny bits are found in the smoke and soot from brush fires, heavy metals, and toxic chemical fumes. But research has consistently shown that $PM_{2.5}$ particles are far more toxic and deadly that the larger particles because they can evade the respiratory system's natural defenses. In California, exposure to these fine air particles is associated with up to 24,000 deaths every year, according to a 2009 study by the California Air Resources Board, the majority of them in highly populated areas such as the San Francisco Bay, the San Joaquin Valley, and the Los Angeles air basins. $PM_{2.5}$ particles penetrate deep inside the lungs, causing constant irritation that diminishes lung capacity and can lead to cancer. Like PM_{10}s, they insinuate themselves inside the walls of blood vessels, which can trigger the formation of the artery-clogging plaques that are the culprits behind strokes and heart attacks. There is evidence that even smaller particles, which are 100 nanometers in diameter, can infiltrate the brain through the nasal passages and erode our cognitive abilities.

Of course, we've known intuitively for years that air pollution promotes respiratory suffering, but perhaps the most striking proof of the link was a 2001 study published in *JAMA, the Journal of the American Medical Association,* in which scientists from the CDC looked at the changes in air quality in Atlanta when it hosted the 1996 Summer Olympic Games. This provided researchers with a rare living laboratory in which to observe what happens when emissions are drastically reduced. The city implemented a number of measures to avert traffic gridlock when 11,000 athletes from 197 countries, along with more than two million visitors, were expected to descend on the Atlanta metropolitan area, which was then home to nearly three million people. The city provided round-the-clock public transportation, adding 1,000 buses to its existing fleet and closing off the downtown area to private cars. Local businesses also changed truck delivery schedules and instituted flexible work schedules and telecommuting options, which greatly eased congestion, especially during the critical early-morning rush hour. This was particularly important because pollutants released by cars in the morning are converted into ozone later in the day. What researchers discovered about daily shifts in air quality during the 17 days of the Games was remarkable: Even though it was humid and sticky and temperatures hovered in the mid-80s, ozone dropped by 27 percent during peak periods, and there was a similar decline in other air pollutants that trigger asthma attacks. The study compared the number of emergency medical visits for asthma during the Games to four-week periods before and after the Olympics. Sure enough, the number of kids with asthma showing up in doctors' offices and in ERs dropped by nearly half.

Indeed, these examples suggest that kids may be the ones who suffer the most. Proportionally, children inhale more airborne toxins per pound of body weight than adults because their metabolisms are higher. To uncover the actual magnitude of this environmental damage, I visited Dr. Kent Pinkerton, a pediatrician at UC Davis who studies these links as director of the university's Center for Health and the Environment. "Children are exposed to a 30-fold greater dose of pollutants compared to adults breathing the same air because they spend most of their time outdoors and, pound for pound, they take in a whole lot more of whatever is in the air," he told me during a recent interview at UC Davis. "Their lungs are still developing and they're not born with a mature immune system, which means they don't filter out toxins entering their lungs as efficiently as adults, making them more susceptible to injury. The first two to six years of life may be critical developmental windows that are the only phase in the life cycle where you can efficiently develop certain features of organ systems. And once you pass these markers, the damage is irreversible."

DEADLY DUST

One component of pollution, diesel fumes, delivers a double whammy for health. The diesel exhaust emitted by factories and big rigs not only damages the lungs, but also makes an excellent transport system for fungal spores, which proliferate in hotter, carbon-enriched environments. They attach themselves like glue to the tiny diesel particles, which scatter them in the wind in a "nasty synergy," to use a phrase coined by the late Dr. Paul Epstein, a pioneer in environmental health at Harvard. The fungi lurking inside the spores can be lethal, which is something Sharon Filip unwittingly discovered when she investigated the possibility of relocating to Arizona from her native Seattle.

Filip, a hypnotherapist, was just tired of the rain. While it made Seattle lush and densely foliated, the weather was dismal for half the year—chilly and overcast or drizzly when it wasn't pouring. Filip yearned for warmer weather where she wouldn't have to bundle up or carry an umbrella and brave driving on rain-slicked highways. In 2001, when one of her sons enrolled at the University of Arizona in Tucson, she thought it was kismet— a sign that it was time to move to a more affordable city where she could bask in year-round sunshine. She flew to Arizona to help him get settled, and toured new homes in some of the upscale planned communities in the hilly northern suburbs above Tucson's downtown, a grid of interlocking strip malls abutting a slab of cookie-cutter subdivisions that had sprouted like weeds during the '90s real estate boom.

Filip was underwhelmed. Even in September, Tucson was scorching hot and dusty, and housing wasn't as cheap as advertised. At home in Seattle one week later, Filip awakened one morning feeling as if she had swallowed rat poison. She was so tired she could barely lift her head off the pillow, her body was racked by coughs, and she had such a pounding headache it felt like her brain was pulsing through her skull. A chest x-ray revealed that her lungs were filled with fluid. Doctors told her she had a severe case of pneumonia and pumped her up with antibiotics. But Sharon continued to deteriorate. "Tears were running down my face from agony," she said.

Two weeks later she was referred to a lung specialist, who immediately diagnosed what was wrong after learning where she'd traveled. Blood tests confirmed that Filip had valley fever. Medically known as coccidioidomycosis, or "cocci" for short, it is a sometimes fatal dust-borne lung disease caused by a soil-dwelling fungus in the mushroom family. The fungi are carried inside tiny spores that form protective cocoons and enter the lungs when inhaled. The fungi flourish in much of the Southwest, in regions with dry weather and sandy soils, like Arizona, Utah, New Mexico, parts of Texas, and California's San Joaquin Valley, which gave the disease its name after a major local outbreak in 1930. Any disturbance of the soil—new construction, crop planting, oil exploration, archeological digging, even earthquakes or high winds that kick up dirt—can stir up a cloud of spores.

About 60 percent of infections are asymptomatic, and those who do get sick may experience flu-like effects—fever, coughs, headaches, and muscle aches—but usually get better within a week or two. In some, the illness can progress to pneumonia. And if the microscopic spores spread beyond the lungs—doctors aren't sure how often that happens—they can attack any organ, and wreak havoc in the bones, liver, spleen, skin, joints, and nervous system, causing meningitis that can be permanently debilitating or even deadly. The elderly, people with chronic illnesses like AIDS or diabetes, young children, and pregnant women, especially those in their third trimester, are more susceptible to becoming gravely ill and dying.

"It took me four months to come back to the world of the living," recalled Filip, who doesn't have any of the risk factors that would make her more vulnerable to infection. For years afterward, she was so worn down she couldn't do the simplest tasks, like heating up a bowl of soup or walking down a flight of stairs. "Even now, more than a decade later," she said, "I still have residual symptoms and my lungs are permanently scarred."

Valley fever has lurked in the deserts of the Southwest for hundreds of years. There have been many notable outbreaks of the disease, including one right after the 1994 Northridge earthquake in Los Angeles and another in 2005 that swept through a prison in Coalinga in Central California and infected more than 900 inmates and prison officials. During World War II, valley fever crippled operations at many army bases in the Southwest and afflicted up to 25 percent of the thousands of soldiers stationed at airfields in the San Joaquin Valley, many of whom were sick for months. Even German prisoners of war held in detention camps in Arizona fell ill and had to be moved after the Nazi government invoked the Geneva Conventions. But in the past decade, the disease's incidence has reached epidemic levels, according to the CDC, which estimates that annually, it sickens about 200,000 people and kills as many as 200, making it deadlier than Lyme disease, influenza, or West Nile virus. The number of infections has more than quadrupled in Arizona and tripled in California, and the actual magnitude could be far higher because many doctors don't recognize the symptoms. "The problem is that doctors aren't sensitized to look for this," said Andrew Comrie, a climatologist and dean of the University of Arizona Graduate College, who investigates the links between climate and disease.

The spike in cases is driven in part by the continued migration to Sunbelt cities like Phoenix, Tucson, Bakersfield, and Fresno, which has led to the construction of vast subdivisions in once-empty deserts. But mainly, the new weather patterns fueled by global warming—the rising thermometer, droughts followed by drenching rains, and parched landscapes with more violent winds kicking up dust clouds that travel for hundreds of miles—are to blame, having created ideal conditions for the transmission of valley fever. The cocci spores, which burrow several layers beneath the soil surface, benefit from the higher temperatures, which encourage fungal growth. What's more, during the scorching summer months, the intense heat kills off surface vegetation so that when the rain does fall, the fungus flourishes without any competition. And because the hotter temperatures burn off the desert grasses, plants, and roots that anchor the soil, more dust is roiled by the churning winds, which often carry the spores for 500 miles or more. As the geographic belt hospitable to the fungus widens with longer, hotter, and drier summers, the number of people stricken could multiply, according to a report by Physicians for Social Responsibility.

In the drought-stricken Southwest—Texas, New Mexico, and Arizona—dust storms reminiscent of the Dust Bowl of the 1930s have become commonplace. They're called *haboobs*, the Arabic word for the intense dusters that can reach several miles high and whip across the Sahara and the Arabian Peninsula at speeds of up to 60 miles per hour. A collision between a cold front or a collapsing thunderstorm and severe, dry weather is what kicks up the ferocious winds and creates a wall of dust in regions with sandy surface soil. In July 2011, Phoenix was enveloped by a 100-mile-wide haboob so fierce that air travel was halted because of the low visibility and high winds, and the sky turned darker than night. Nearly 10,000 residents lost power, and levels of particulate matter at one monitoring site hit an hourly average of 5,000 micrograms per cubic meter, about 300 times what is deemed safe by the EPA. Local hospitals braced for an onslaught of people with asthma, bronchitis, and valley fever.

The fierce, drought-driven dust storms that propel fungal spores aloft "are now just as serious [a health hazard] as traffic and industrial air pollution," Laurence Barrie, chief researcher at the World Meteorological Organization in Geneva, said recently. "The minute particles act like urban smog or acid rain. They can penetrate deep into the human body." The size of these dust storms has astonished scientists, and their range will only increase as the changing weather bakes the moisture out of the soil and whips up higher-velocity winds. Satellite pictures reveal that some clouds originating in Africa, which is gripped by a brutal dry spell, are as big as the continental United States and can unleash a million tons of soil into the atmosphere, while dust blown from storms that start in the Gobi and Taklamakan Deserts in Asia often contribute to smog over Los Angeles and poor air quality along the West Coast. These dust storms carry not only minute particles of fungal spores, but also potentially harmful bacteria and viruses. Saharan storms are believed to have spread deadly bacterial meningitis spores throughout Central Africa, where up to 250,000 people are sickened by it every year. Dust clouds are also thought to be responsible for the transmission of influenza, SARS, and other infectious respiratory ills. And as climate-change-induced desertification increases in the Middle East, in Africa, and here in the United States, said UC Davis's Kent Pinkerton, "it's going to create more of these sandstorms that transport particles such long distances."

Hoodlum Weeds

The pollens released into the air by the flowering plants and trees that cause hay fever and provoke asthma attacks are also changing in response to altered climate patterns. Pollen seasons now start up to a month earlier and end later because of warmer temperatures. They sometimes stretch from February until June in the nation's southern tier instead of the more typical March through May. Hotter weather coupled with higher concentrations of carbon dioxide in the air prompt allergens like ragweed to grow larger and produce more pollen. Even more disturbing, the carbon-enriched pollen—carbon dioxide, after all, is a driver of photosynthesis—is now more noxious and damaging because it contains more of the chemicals that cause allergic reactions. That was the conclusion of a groundbreaking study from a USDA research team led by plant physiologist Lewis Ziska, demonstrating yet again how the changing ecological balance of nature threatens our health.

The lanky, outgoing Ziska, who has a thinning thatch of sandy hair and a moustache, has spent much of his career looking at how climate change will affect one of humanity's enemies: weeds. These hardy survivors are the bane of gardeners' and farmers' existence, costing agribusiness $33 billion annually in both lost crops and control efforts. They almost taunt us with their ability to thrive despite our best efforts to eradicate them with toxic sprays or by ripping them out of the ground.

At his lab at the USDA campus in suburban Maryland, Ziska wanted to know what effects a warming planet would have on plant biology and how growth would be affected by the increased concentrations of atmospheric carbon dioxide and higher temperatures that climate models are projecting in the next 50 years. But the study design was dictated as much by ingenuity and economics as by the rules of scientific exploration. Severe budget cuts during the Bush administration had decimated his lab, reducing the number of scientists from 10 in the waning days of the Clinton White House to only 3 researchers studying climate change. Erecting the costly facilities to create airtight, climate-controlled growth chambers—essentially, aluminum boxes about the size of meat lockers—was out of the question. "The CO_2 alone would have cost a million dollars a year," Ziska told me.

But Ziska hit upon a novel solution that capitalized on the intense heat island effect of neighboring Baltimore, where pavement and buildings radiate the solar energy they absorb and exhaust fumes from cars and big rigs and ships in the harbor clog the air with smog, forming a CO_2 dome over the inner city. Downtown temperatures were registering 3°F to 4°F warmer than in the suburbs, and CO_2 concentrations were 440 to 450 ppm, which was considerably higher than the 393 ppm average of 2012 and came close to the conditions moderate climate change scenarios are predicting for the planet as a whole by the middle of this century. "We were looking for a cheap way of getting at future climate," said Ziska. "We figured the urban area might be a pretty good surrogate for that."

Ziska planted weeds—using seeds from about 30 different types that he culled from an organic farm—on a barren patch of abandoned land next to the city's Inner Harbor to see how they would fare in the hotter, more hostile CO_2-enriched environment. For comparison, he planted the same weeds at two other sites: an organic farm in western Maryland, where growing conditions approximated the current climate, and a park in a suburb of Baltimore, which presented circumstances between these two extremes. They made sure the seeds and the soil composition of the plant beds were the same so the only variables would be CO_2 concentration and temperature.

What happened over the next five growing seasons was astonishing. Turbocharged by the warmer temperatures and higher levels of carbon dioxide, the downtown weeds behaved like they had been doused with some magic fertilizer, soaring up to 20 feet, quadruple the size of their brethren on the farm. "In one sense, it's amazing because they are the ultimate adapters," said Ziska, who admitted that he felt a certain grudging admiration for the tenacious weeds. "Even with the most extreme ecological disturbances, the first thing you'll see coming back are the weeds."

Even worse, the chemical makeup of the CO_2-enhanced weeds seemed to change, with the pollen they generated not only being four times as plentiful as that of their rural cousins, but also more allergenic, because they have more of the protein responsible for allergic reactions. While carbon dioxide is a basic greenhouse gas, Ziska pointed out, it's also a food source for plants, which convert it into sugars and carbohydrates. But not all plants respond the same way, and noxious weeds—as well as vines like poison ivy and kudzu—react much more strongly to higher CO_2 than other types of plants do, he said. As a consequence, we see not only more growth, but also more virulent chemicals within the plants.

These findings have dire implications for public health and also for farmers, who are already battling infestations of seemingly indestructible superweeds. A more recent study that Ziska collaborated on with 20 other mostly

US–based researchers had even more bad news for the nation's 50 million hay fever sufferers. In analyzing US and Canadian data on ragweed and daily temperatures at 10 different latitudes, they found that pollen seasons were 27 days longer than they had been in 1995 in the northernmost areas, where temperatures are rising the fastest in response to climate change, and the lengthening of the pollen season was directly correlated with the hotter weather because the frost is melting earlier in the year.

Putting all these pieces together—longer pollen seasons, more powerful weeds producing additional and more noxious pollen, and the spreading geographical footprint of pollen seasons—means that climbing temperatures, increased smog, and record pollen counts are creating a combustible mixture that is sparking an explosion of new allergy and asthma patients. "When we look at climate change, we think of it as this esoteric thing that will happen to our grandkids, but the fact is, this is already occurring," said Ziska, who is deeply frustrated that so few are paying attention to the disastrous changes in the ecosystem. "It's the hardest thing in the world—it's like seeing the train coming and not being able to speak English to warn people. Or a 'black swan' event—a transformative event that no one anticipated but in hindsight, we should have seen coming because all the signs were there that it was going to happen."

SIMPLE FIXES CAN CLEAR THE AIR

We live in a global ecosystem, and what goes on in remote regions can have a huge impact on worldwide pollution. By the same token, even seemingly small changes in curbing greenhouse gas emissions can not only reduce harmful pollutants, but also help to slow climate change. Intriguingly, the second-biggest contributor to global warming—surpassing even the emissions generated by farming, industry, and powering our homes—is from a routine daily activity that is virtually unknown to inhabitants of industrialized nations: burning charcoal, animal dung, or wood to cook food. More than three billion people, almost half the world's population, rely on these solid fuels for heating and for preparing meals.

The atmospheric changes triggered by these cooking fires affect us as well. The smoke contains black carbon, CO_2, methane, and other pollutants that play key roles in climate change, and dumps billions of tons of black carbon—the second-most-deadly greenhouse gas, behind only CO_2—into the atmosphere every year. A component of soot, black carbon absorbs solar radiation and is partly responsible for the accelerated melting of polar and alpine glaciers. And this toll doesn't include the collateral environmental damage caused by the use of firewood, which has left vast swaths of the countryside in rural Africa and Asia denuded and caused severe erosion where dirt has washed away. Cutting down trees also eliminates an important carbon-absorbing sink for greenhouse gases, which means more destructive pollutants remain in the atmosphere. Given the explosive population growth on the African continent and the rapid rate of deforestation, experts predict firewood supplies will be mostly depleted within the next quarter century.

It's a little-reported issue that is finally hitting media radar screens: Millions of women in developing countries spend many hours every day sitting in poorly ventilated rooms cooking food over small fires. Being in one of these makeshift kitchens, with air thick with soot and walls blackened from the tar of the open fires, one reporter noted, is "like sitting inside a smoker's lung." Breathing that polluted air day in and day out ravages the lungs, leading to cancer and heart and respiratory diseases, and causing more than two million premature deaths a year, mainly among women and their children, according to United Nations estimates.

In June 2010, US Secretary of State Hillary Rodham Clinton, speaking at the Working Women's Forum in Chennai, India, drew the industrialized world's attention to the dangers of inhaling these toxic fumes and pledged $50 million in seed money to the Global Alliance for Clean Cookstoves, which will supply 100 million women in Africa with fuel-efficient indoor stoves. Installing a low-cost stove in millions of homes in India, Africa, and Latin America can not only save more lives than far-costlier vaccination campaigns, but also greatly reduce the spewing of hazardous substances into the atmosphere. This initiative is the result, at least in part, of painstaking, immaculately designed studies conducted in a remote Guatemalan village by a research team led by Kirk Smith, a professor of global environmental health at the University of California, Berkeley, who has spent the past three decades studying the health effects of indoor pollution.

And it underscores the fact that relatively inexpensive remedies can help clean up the air.

One temperate June day, Smith sat down with me in his cluttered office in a 1960s-style glass-and-steel high-rise at the edge of the UC Berkeley campus and talked for an hour about how this dirty air impacts the health of impoverished people around the world and how a simple fix can pay off in multiple ways. A compactly built man, he was casually dressed in jeans and a long-sleeve madras shirt. Now in his 60s, his moustache, shock of curly gray hair, heavy-lidded eyes, and genial, dry wit at first made him seem more like an aging hippie who had wandered in from nearby Telegraph Avenue than a world-class scientist with bookcases full of awards. The school year was over and the tree-lined campus, with its magisterial granite and art deco–style buildings, was deserted. But the scientist's office was a beehive of activity, with a scrim of graduate students crowded outside his door and a crammed schedule of faculty meetings. Perched on a windowsill was a gleaming steel prototype stove, a little bigger than a gallon can of paint, that a Dutch appliance maker had fashioned to prove it could be mass-produced for less than $50 apiece.

A physicist by training, Smith started his career working on nuclear energy problems. But he soon realized that even an accident the size of Chernobyl once a month wouldn't come close to the magnitude of the health effects of air pollution, with a million people dying due to the by-products of coal combustion and two million killed by the effects of indoor air pollution every year. If he wanted to improve people's lives in a meaningful way, he knew, he needed to follow the risks.

They looked at a number of sites in the developing world—Turkey, Nepal, India, Ethiopia, Kenya, South Africa, Peru, and Mexico, and finally settled on Guatemala. But it took another 17 years of applying for grants before the National Institutes of Health finally gave him $500,000 in 2001. The following year, Smith, along with a group of students, researchers, and Guatemalan health experts, began making the arduous treks to spend one week out of every month in San Lorenzo, a tiny village nearly 9,000 feet above sea level in the western highlands of Guatemala, where natives had no electricity and many spoke only the indigenous language.

There, the researchers recruited more than 500 local families with either a pregnant mother or an infant. Each household was randomly assigned to receive a cookstove with a chimney that vented smoke or to continue cooking over open fires. Electronic sensors and transmitters were attached to the walls and the villagers' clothes to measure particle emissions and track the amount of carbon monoxide they were exposed to every day. Once a week, the data from the sensors were downloaded and the families were examined by Smith's medical team. This scrupulous recording of how much smoke residents were exposed to every day and their medical conditions yielded a wealth of compelling evidence.

They were able to show a direct correlation between smoke and disease, with the dirty air causing far higher rates of pneumonia and adverse birth outcomes in children and cataracts, tuberculosis, heart disease, and chronic lung disease in women. In women in their third trimester of pregnancy, constant exposure to the carbon monoxide in woodsmoke had serious health consequences for their unborn children. These babies had birth deformities such as cleft palate; did poorly on cognitive and neurodevelopmental tests at ages six and seven, showing severe deficits in fine motor skills, visual perception, and motor memory; and exhibited an average decline in IQ of about five points compared to unexposed Guatemalan children of their age. These types of impairments can curtail an individual's lifetime earnings, their educational attainment, and their overall success in life. What's more, children who inhaled the least smoke were up to 85 percent less likely to contract severe pneumonia than those who inhaled the most. "The amount of smoke exposure babies were getting from the open wood-fire stoves is comparable to smoking three to five cigarettes a day," said Smith.

While the results seem intuitive—spending long hours breathing in smoke in a poorly ventilated room is bound to make anyone sick—actually nailing down a direct cause and effect with the kind of irrefutable hard data that satisfy picky scientists is no mean feat. Since half the world's population cooks this way, the health impacts of this exposure, Smith told me, are larger than those of any other environmental risk except contaminated water supplies.

Despite the magnitude of the problem and his persuasive data, Smith spent many fruitless years trying to get major foundations, which are deluged by funding requests for equally worthy projects, interested in sponsoring a stove-buying initiative. "It was very frustrating," he sighed. "We'd go to an air pollution meeting and present our results, and people were astounded: 'Good lord, this is a thousand times worse than cities.' And then I'd go across the street to the international health meeting, and they'd say, 'Well, Mr. Smith, you have a serious risk factor. But, you know, I've got'— and this is the number in India—'$7 per year per capita to spend on public health. Are you saying we take money away from antibiotics and vaccines for improved stoves?' So that's where we're stuck. The kind of evidence we had 30 years ago would be enough to call out the National Guard in this country, but isn't enough yet, really, to trigger the response, because, well, we aren't even vaccinating the kids, not even getting them the antibiotics every kid ought to have. A poor rural woman in a developing country is the most politically disenfranchised, about the bottom of the totem pole, and then her children are even lower.

"My students sometimes get discouraged," he continued, barely pausing to take a breath. "And I tell them, policy is lumpy. Nobody pays any attention for years, and then, all of a sudden, they call you up Monday morning and they say, 'We need to know the answer by Tuesday afternoon.' And if you don't already have it in the literature, you can't do that. You don't know when that Monday morning is going to happen. But you've still got to get it."

Smith's Monday morning finally arrived—and he certainly couldn't have a better champion than Hillary Clinton. Of course, this is only the first step in a long process, but it's a real beginning.

SIMPLE JUSTICE

Local grassroots groups have also been successful in pushing polluters to clean up their communities and compelling government agencies to protect residents from the consequences of climate change. They're part of the environmental justice movement, a crusade that grew out of the recognition that it is mainly the poor and people of color who are forced, by circumstance, finances, lack of political power, and what activists call "environmental apartheid" callously targeting the disenfranchised, to live and work in some of the nation's dirtiest environments.

Luis Cabrales, who worked for many years as a campaign director for the Coalition for Clean Air, a venerable green group that was instrumental in the passage of California's historic vehicle Smog Check Program in the early 1970s, has spent most of his adult life fighting for environmental justice. A slight man with a square jaw and the broad-shouldered build of a wrestler, Cabrales possesses the easygoing charm and dogged persistence of a natural-born organizer. The son of illegal immigrants from Sonora, Mexico, he graduated cum laude from California State University, Los Angeles, even though he worked full-time to support his mother and youngest brother; his father had died of lung cancer when Cabrales was barely 11. His college years had been haunted by the constant threat of deportment because he was undocumented. "An illegal alien," recalled Cabrales, the words obviously still carrying a sting. "I worried all the time and I couldn't apply for grants or financial aid."

There have long been strands of environmentalism in the civil rights struggle, Cabrales tells me during a lengthy interview in the coalition's offices in a high-rise in downtown Los Angeles: Rosa Parks's refusal to move to the back of the bus stemmed from a deep need for adequate public transportation, he points out, while César Chávez pushed to protect farmworkers from the harmful pesticides that were killing them in California's Central Valley. But it wasn't until the late 1970s that people started connecting the dots and realized that African Americans resisting attempts to establish garbage dumps in their neighborhoods were essentially fighting the same fight as grassroots activists protecting endangered species. "Better housing, clean water, clean air, safe schools," said Cabrales, who became a citizen when he married an American, "these all had to do with the environment."

The turning point came in September 1982, many observers believe, when dump trucks carrying hazardous waste laced with toxic PCBs first arrived in Warren County, North Carolina. They were headed for a landfill in the town of Afton, a poor rural community that was overwhelmingly black. Many locals were infuriated that authorities had ignored their concerns about poisonous chemicals seeping into their groundwater. They stopped the trucks by lying down on the roadways leading to the dump, igniting six weeks of protests. More than 500 people were arrested, which generated national media attention—and launched a potent and enduring grassroots crusade.

In the decades since, the environmental justice movement has grown from a scrappy army of the poor and disenfranchised into a powerful coalition that united the mostly white and college-educated traditional environmentalists, trade unionists, blue-collar workers, and residents of some of the nation's most contaminated neighborhoods. It has used its growing political clout and numbers to take its place in the corridors of power. "We've changed the paradigm into a green-blue coalition," said Cabrales. "Years of environmental consciousness raising are really bearing fruit, and environmentalists are no longer seen as weed-smoking, sandal-wearing tree huggers. Over the last 10 years, the change has been dramatic because a new generation of environmental advocates is now holding positions of power in the legislature, on the ports commission, and on the air resources board."

It's a remarkable sea change that has seen someone like Fabian Nuñez of East Los Angeles, 1 of 12 children born to a gardener and a maid from Tijuana, ascend to the speakership of the California Assembly and coauthor the nation's most stringent measures to alleviate global warming. The landmark package of laws is considered former governor Arnold Schwarzenegger's signature legislation and has become a model for the rest of the world. "He was a gardener when he was a kid and pushed around a lawn mower helping his dad," said Cabrales with no small measure of pride. "Now he's helped create the strongest pollution-control standards in the world—a Mexican immigrant."

In the past decade, the Coalition for Clean Air has waged a successful campaign to begin cleaning up the deadly diesel pollution emanating from California ports. Working in tandem with such groups as the Natural Resources Defense Council and the Coalition for a Safe Environment, along with the Teamsters' and the longshoremens' unions and local activists like Martha Cota, they've done extensive lobbying in Sacramento and community organizing to push for enforcement of the California Environmental Quality Act and compel the port of Los Angeles to initiate a suite of pollution-reduction strategies that have cleaned up the air. These included the Clean Truck

Program, which replaced the port's fleet of old, dirty diesel trucks with EPA-compliant vehicles equipped with particle filters, and electrifying the ports so ships can just plug in to power their infrastructure while at the dock. A single ship idling in the port—and they normally sit at anchor for two to three days while they're being unloaded—emits more pollution than five diesel school buses in an entire year. Turning off the engines and using electrical power can cut as much as three tons of smog-inducing emissions from each ship. Still, while pollution has been reduced by about 70 percent, air quality remains a serious problem. "But at least we've made a start," said Cabrales, and he added, flashing a toothy smile, "we now have a seat at the table."

CHAPTER THREE

THE HOT ZONE

"No one saw it coming."

George Luber was describing one of the deadliest climate disasters in history, a lethal heat wave that scorched Europe in 2003, killing a staggering 52,000 people across the continent. The highest death toll was in France, where 15,000 died, most of them elderly. Temperatures hovered in the low 100s for two weeks and failed to dip much at night. "The nighttime heat is especially dangerous, and that's what drives mortality because people can't cool off," Luber recalled in a series of interviews conducted over the phone and in person. "It was so hot in France that they came very close to having their nuclear power plants shut down. The temperatures in the rivers got so high that they couldn't cool down the reactors."

Luber was a young epidemiologist with the CDC in Atlanta when the head of his division got a call from Paris doctors and public health officials that August. They had come to the stark realization that they were in the midst of a crisis for which they were disastrously unprepared. Northern European countries like France aren't accustomed to sizzling summers—daytime temperatures rarely climb higher than the mid-70s, particularly in northern regions, and nights are even cooler, so few homes have air-conditioning. Unfamiliar with sweltering weather, Europeans not only didn't know what to do to cool off, but also didn't understand how imperative it is to do so, or that sustained exposure to high heat can be fatal. And the hot weather couldn't have happened at a worse time— August, when most of the continent is on holiday, including physicians and politicians, and skeleton crews are left to staff rescue teams and hospitals. Adult children were out of touch with their parents, while elderly relatives who were unable to travel were parked in nursing homes.

Luber was sent to Paris—it was the first time he parachuted into a crisis for the CDC—and he arrived just as the heat wave broke. Governmental authorities appeared to be in complete denial about the gravity of the emergency—the health minister initially refused to cut short his vacation, while president Jacques Chirac remained in Canada on holiday for the duration—and the situation had descended into utter chaos. "There was an absolute sense of disbelief among epidemiologists about the magnitude of the problem—and several top health officials eventually lost their jobs," recalled Luber, who as the days wore on became increasingly frustrated by their apparently complacent response to the perilous weather.

In an average year, heat waves kill more people than all other natural disasters combined, claiming about 1,300 lives here in the United States. They are silent killers, Luber noted, "natural disasters that do not leave a trail of destruction in their wake." While hurricanes, fires, and floods wreak long-lasting material havoc—impassable streets, flattened buildings, power outages, demolished infrastructure—most of the visible evidence of heat-related damage vanishes the minute temperatures drop, so the extent of the carnage is easy to forget.

Often, too, doctors don't attribute the deaths directly to the heat, even when heatstroke is the most obvious cause, for a number of reasons: The victims may be elderly or have other illnesses that could have played a role, or the bodies have cooled down by the time they're discovered. Consequently, we don't have a real sense of just how catastrophic hot spells can be or the true magnitude of the fatalities. "One of the tricky things about heat waves is that you would only notice the elevated number of deaths if you were looking for them, and then you suddenly find these huge spikes of deaths that you've never seen before," said Eric Klinenberg, a sociologist at New York University and author of *Heat Wave: A Social Autopsy of Disaster in Chicago.* Time is of the essence, too: After 48 hours of constant exposure to temperatures in excess of 90°F, the body's defenses start to break down. Consequently, the swiftness of the public health system's response to heat-related illnesses can literally mean the difference between life and death;

even a 24-hour delay can translate into a much higher rate of mortality.

Public health officials received a blistering lesson in this hard truth when Chicago was gripped for more than a week by a fierce heat wave in July of 1995. Temperatures hovered in the low 100s, and the heat was intensified by unusually high humidity—dew points in the low 80s—that made the sweltering city feel like the tropics. Temperatures hit 106°F at their peak, but the heat index—how hot it feels to your body, analogous to winter's wind chill factor—climbed to 126°F. Thousands of Chicagoans were sickened by serious heat-related illnesses, and ambulance services were so overwhelmed that the city dispatched fire trucks to respond to 3,900 calls.

Heat cramps, normally the first sign that the body is having trouble coping with high temperatures, are typified by severe cramping in the legs or stomach, exhaustion, or dizziness. These can usually be controlled by drinking plenty of fluids and going to a cool place. Heat exhaustion is far more serious, and it is caused by the body losing too many fluids through heavy sweating from either vigorous exercise or being in a hot, humid place. Sufferers experience headaches, fatigue, nausea, and vomiting; they perspire profusely and their body temperatures rise. If they aren't treated quickly, their condition can swiftly escalate to heatstroke. This is life threatening because the body's natural cooling system fails and they stop sweating, causing their core body temperatures to shoot up to 106°F or higher. That interferes with the electrical system that controls the heart's rhythm, causing a weak or rapid pulse, and can also trigger confusion, seizures, coma, brain damage, and even death.

In the Windy City that July, the temperatures were so high that roads buckled, train rails warped, a power generator burst into flames, and bridges had to be hosed down with water to prevent them from locking when the plates expanded. Paramedics couldn't keep up with emergency calls, and many worked 28-hour shifts in 100°F heat to help as many as possible, while hospital emergency rooms were so crowded that 23 of them temporarily stopped accepting patients. Three days into the heat wave, the 222 bays in the city's morgue were full and the county had to rent nine refrigerated meat trailers to store hundreds of additional corpses. Ultimately, more than 700 people perished, an additional 3,300 desperately ill people were rushed to emergency rooms, and many of those who survived suffered from the lingering after-effects of heatstroke, such as worsening of heart disease or kidney failure, according to subsequent research. Another troubling study confirmed that nearly half of the heatstroke patients treated in the intensive care unit died within the year, while one-third were left so severely impaired that they were unable to live independently thereafter. Because emergency rooms were overwhelmed, the same study found, few patients received optimal treatment, which meant that many could have avoided profound long-term health consequences.

Yet while the crisis was unfolding, civic officials in Chicago failed to take action to stave off fatalities: Emergency managers didn't request reinforcements to aid overworked paramedics and ambulance drivers, and mayor Richard M. Daley callously downplayed the deadly consequences of the blistering heat. He dismissed the heat wave as an act of God, raised questions as to whether the crisis was "really real," and forbade the health department to release the number of dead. Daley and many members of his cabinet took off for summer vacations before the heat wave started even though meteorologists had warned that scorching weather was on the way, and Daley returned two days into the crisis only because "dead bodies began piling up at the morgue," Klinenberg told me in a recent interview. "The real shame of the Chicago event was the political refusal to take it seriously. Public relations concerns trumped public health concerns during a crisis."

It seemed to George Luber as if the same governmental negligence was repeating itself in Paris. There, however, the loss of life would reach cataclysmic proportions because of the scarcity of air-conditioning and of public parks in which to cool down in the ancient, highly urbanized city, which is honeycombed with narrow alleyways that intensify high temperatures. Based on what had happened in Chicago, Luber was keenly aware that the first 48 hours of a heat wave are a critical lifesaving window. Heatstroke can be resolved with immediate medical attention. But once the body's systems start shutting down, the cascade of damage is irreversible. "They might survive the first day, but as the days wear on, they become weaker and weaker," Luber said. "The resistance to heat gradually erodes, and [by] the second or third day, people are no longer able to cope. There is a lag between the spike in temperature and death. Once the deaths start piling up, it's too late to mount an effective response."

But French authorities made no attempt to locate the people who were most vulnerable to the effects of the suffocating hot spell—the old, the poor, and the infirm, particularly those who lived isolated from their neighbors and networks of family, friends, and social service agencies. "We kept telling health officials that they needed to evacuate the nursing homes and get people to temporary cooling shelters, but that sense of urgency was not there," recalled Luber, whose own experience with heatstroke is a testament to the power of quick intervention and the vigilance of others: As a young competitive tennis player, he had been stricken with heat exhaustion in the middle of a match. "I was walking in circles and started vomiting," the Florida native recalled. "I was taken off the court to a hospital, where I spent three days. I just totally lost the ability to function. If I hadn't had eyes on me, I could have progressed to life-threatening heatstroke and even died."

Within those crucial first few days in Europe, thousands fell ill, but the window for rescuing many of them had already slammed shut. By the time ambulance crews delivered some extremely sick victims to the hospital, they were too dehydrated and feeble to be revived. A handful died en route. Within an eight-day period, more than 30,000 across the continent lost their lives. In France, doctors in swamped hospital emergency rooms sounded alarms about the swelling tide of death, and undertakers at mortuaries alerted city bureaucrats that they were seeing an exceptional number of fatalities, but their reports were ignored. Paris's 450-person-capacity morgue rapidly filled with corpses, and funeral homes were equally overwhelmed. The conditions reached a crisis during the second week of August. The overcrowding at the morgue finally got the attention of the Paris police and those running the Ministry of the Interior, who belatedly cut short the vacations of emergency and health care personnel 10 days into the crisis. But by then, it was too late to stop the loss of life.

Undertakers were forced to rent a refrigerated warehouse outside Paris because they didn't have enough space at their own facilities. Many victims' bodies remained unidentified for weeks because their families were on vacation. By early September, nearly 60 still sitting unclaimed in lockers were buried in unmarked graves. Thousands of people in the heart of the city died alone, without air-conditioning or even adequate ventilation, in tiny quarters no bigger than walk-in closets on the highest floors of dilapidated Paris apartment houses. Police and firefighters made ghastly discoveries of decomposing bodies that had been baked in ovenlike top-floor rooms ventilated only by skylights. "It's unfortunate that the City of Chicago didn't do more work to call attention to what happened there in 1995, because the pattern of what happened and who died were just stunningly similar," said Klinenberg of the French disaster.

The real tragedy in all of this is that none of these people had to die. Deadly heat waves aren't like flash floods or tornadoes that can hit with little warning. They're easy to predict, and some relatively simple preparations can prevent human suffering and save thousands of lives. In one retirement home in the Paris suburb of Cachan, the facility's director had thoughtfully implemented a series of risk prevention measures ranging from medical procedures to ice packs, wet towels, fluid administration, and routine airing out of rooms to make sure the 135 permanent residents were hydrated, cool, and cared for in the stifling weather. While the surrounding region experienced the sharpest spike in excess deaths in France (171 percent), this facility lost only one severely handicapped patient.

But what happened in most of the chronically understaffed and underfunded eldercare facilities was an entirely different story. About half the fatalities occurred in France's retirement homes, where care came up disastrously short. "A lot of the nursing homes were older and didn't have enough bathing facilities," recalled Luber. "We needed to get these folks into tubs and showers, but they didn't have the capacity or personnel and we couldn't cycle them through fast enough to make that an effective response. That was a systemic failure. If they had come to terms with the true magnitude of the risk, they would have certainly had the infrastructure to prevent a lot of deaths."

On September 10, a front-page story in *Le Monde*, France's leading national newspaper, finally acknowledged the enormity of the disaster: The summer of 2003 had been the deadliest in France since World War II.

In fact, it had been the hottest summer in Europe since 1540, when Henry VIII sat on the English throne, according to data collected from ice cores and tree rings. But in the future, heat waves like the 2003 scorcher are expected to become the norm, with average summer temperatures exceeding the highest temperatures ever recorded, according to a landmark study conducted by a team of British climatologists at the Hadley Centre for Climate Prediction and Research and published in 2004 in the journal *Nature*. Heat-trapping pollution doubled the likelihood of the 2003 heat wave, the researchers concluded in the first-ever calculation of how much human actions contributed to the risk of a weather event. By the 2040s, if we continue dumping huge amounts of CO_2 into the atmosphere, the study predicted, more than half of Europe's summers will be warmer than that of 2003.

What happened in France gives us a glimpse of what's on the horizon here, as well—witness the summer of 2012's freak weather patterns. Colorado and New Mexico experienced the worst wildfires in both states' history, with blazes incinerating thousands of acres of bone-dry land and hundreds of homes and forcing the evacuations of thousands. Sizzling heat waves set more than 3,000 daily temperature records in June alone, with huge swaths of the eastern seaboard and the Midwest wilting under triple-digit temperatures virtually unprecedented so early in the summer. The scorching heat sparked the fierce derecho described in the Introduction, which left 3.7 million households in the eastern half of the country without power—some for up to a week—just when they needed air-conditioning the most.

Over the next century, the thermostat is predicted to ratchet up another 2°F to 11°F on average, and more than that in some regions, according to a June 2009 report from the US Global Change Research Program (GCRP), which could mean sweltering summer temperatures in the 110°F to 120°F range becoming commonplace across the country. By 2090, the thermostat may even push past 90°F some 120 days a year in Kansas, and more than half the year in much of Florida and Texas. The Sunbelt will endure more than two months each year of 100°F weather, while in the southeastern Atlantic and Gulf states, climate projection models predict average increases of more than 25°F year-round.

The rapid growth of cities—by 2030, about 60 percent of the projected global population of 8.3 billion will be urban dwellers—will further amplify the effects of rising temperatures because of the urban heat island effect, in which concrete, asphalt, and blacktop retain and radiate heat. This hothouse effect is intensified by a lack of the ventilation that can cool the urban canyons created by tall buildings. "Los Angeles already spends about 25 percent of its energy budget just accommodating for the urban heat island effects," said George Luber.

Chicago's sizzling weather in 1995 was only a prelude, according to the GCRP report. Under a business-as-usual scenario, heat-related deaths in Chicago alone could rise 10-fold by the end of the century. The frequency of extreme heat events—when the temperature rises above 100°F—may triple in many cities over the remainder of this century, with the aggregate death toll climbing to more than 150,000 Americans, according to a 2012 study by University of Miami scientists. Urban centers in the nation's midsection, like Louisville, Detroit, and Cleveland, where residents tend to be older and aren't acclimated to oppressively hot days, will be hit the hardest, with up to 19,000 additional fatalities in each metropolis. "The Midwest won't turn into a Phoenix, but it will be much hotter and drier," said Laurence Kalkstein, the study's lead author and head of the Synoptic Climatology Lab at the University of Miami. "These cities are actually more problematic than the tropics near the equator because temperature increases are greater as you move toward the poles."

The actual magnitude of the fatalities from a heat wave tends to be seriously underreported because there is no national benchmark for determining whether a death was caused by the heat, and local health officials often have trouble deciding whether it resulted from the high temperatures, natural causes, or chronic illnesses exacerbated by the heat. "The elderly and the very young are the most vulnerable; they're not able to cool down their bodies efficiently because they don't sweat as much," Luber said. "This increases the risks of life-threatening consequences when their body temperature rises."

Kalkstein told me that coroners normally list one of these underlying conditions as the cause of death rather than attributing it to the weather. "Medical examiners have strict rules as to what they can call 'heat-related death'—the body has to have certain core temperatures and other specific features before they can attribute it to the heat," he said. "But what they miss are all the heart attacks, strokes, and respiratory deaths that were brought on by the high temperatures, even though there were twice as many [as usual] during the heat wave. In general, we see underestimates by a factor of two or three. We never use the medical examiner's numbers. Instead, we use the spike in deaths that occurs during the periods when these air masses overhead are creating havoc."

Heat strains the heart because the body cools off by sweating, which depletes fluids and decreases blood volume. It also dilates blood vessels, speeding up the heart rate and lowering blood pressure. Consequently, the heart has to work harder to deliver blood to muscles and vital organs. But if we lose too much fluid and become dehydrated, our body temperature rises, causing problems in the nervous system that result in dizziness, headaches, and, in more severe cases, seizures. We lose electrolytes—compounds that perform functions vital for life, such as keeping the heartbeat regular, by regulating the interaction between fluids and our cells—in our sweat and urine. The net result is that losing too much of your body's fluid in extreme heat conditions can cause brain and heart damage, possibly leading to strokes and heart attacks.

In places like Miami, where residents' bodies have, over time, become more biologically tolerant of hot weather, the incidences of heatstroke and heat-related deaths are relatively low. People who live in the tropics are acclimated to hot weather because they lose less salt in their sweat and start sweating sooner, which allows their hearts to pump more blood with each stroke. But there may be a limit to how much we can adapt, according a disturbing 2010 study done by Steven Sherwood, the atmospheric physicist who codirects the Climate Change Research Center at the University of New South Wales (see the Introduction for more detail) and Matthew Huber, director of the Purdue Climate Change Research Center at Purdue University. In the not too distant future—"even within the lifetimes of babies born today," Sherwood told me—we might reach a warming threshold beyond which we can't physiologically tolerate the heat. Summer-day bike rides with the kids, early-morning runs by the lake, and even walking more than a block would become taboo, relics of a bygone era of chillier temps.

How hot does it have to get before we become virtual prisoners inside artificially cooled chambers? That was the question these two scientists attempted to answer.

The wild card here, they pointed out, is the humidity. When humidity is low, we can perspire and cool off as it evaporates, but when sizzling temperatures combine with high concentrations of moisture in the air, we're in trouble. The human body maintains a core temperature (the temperature around our vital organs) of about 98.6°F, but the work our bodies do to breathe, digest, and so on generates about 100 watts of heat when we're at rest, and that goes up to 500 watts when we're digging ditches or running marathons. To offset that extra heat, the body very efficiently keeps the skin temperature slightly below that of the core, at about 95°F, so heat is dispersed through the skin. This delicately calibrated thermoregulation is what keeps us functioning. But when the air temperature and humidity level both rise high enough, we can't bring down the core temperature because the outside air is hotter than it is and too saturated to allow evaporation from the skin—and therefore cooling—to occur. When the humidity is above 70 percent, sweat doesn't evaporate and instead just drips off the skin. Without the cooling effect of evaporation, the core temperature rises. Under these conditions, once the core temp hits 104°F, hyperthermia sets in and organ damage occurs. Anything higher than that is deadly.

At what point does it become just too darned hot? Sherwood considered this question when we had dinner one drizzly April evening after a climate change conference in Cairns, Australia. "Twelve degrees [Fahrenheit] above preindustrial levels is where it becomes physically impossible to survive heat waves in many places," he said. "But how soon before it gets uncomfortable enough given all the other things, like wearing clothing and having to work during the day? I suspect that at seven degrees we'll see mass migrations away from some of the most humid places on Earth—the Amazon, parts of India, northern Australia, and other regions."

When I asked Sherwood why we couldn't just all turn on the air-conditioning, I could almost hear the synapses in his brain crackling with frustration. Air-conditioning, after all, has been a savior in sweltering temps and is considered one of the great inventions of the 20th century, especially for residents of Sunbelt states like Arizona and Florida that are virtually uninhabitable for much of the year without it. About 85 percent of Americans now have these cooling units in their homes. Of course, there are numerous downsides to living inside climate-controlled bubbles, because they create self-perpetuating and ultimately unsustainable feedback loops. Powering the AC requires vast amounts of electricity: Americans use as much energy to keep those units humming as the entire continent of Africa uses for all of its needs. But using more energy means burning more fossil fuels, which results in more greenhouse gas emissions. This in turn ratchets up temperatures even more, which prompts us to turn on the air-conditioning more often.

"It's a totally unrealistic solution—you're never going to air-condition a billion people in Indonesia or South America, it's just not going to happen," he told me. "And it won't protect livestock or people who work outside. Plus, air conditioners require huge amounts of energy, and the price of energy is only going to go up. Billions of people in developing countries can't afford air-conditioning right now. In 30 years, I would question how many people even in the United States, in places like Texas, Florida, and Louisiana, which are hot and humid, could afford air-conditioning. At some point, power failures would be life threatening."

That nightmare scenario could happen sooner than we imagine, according to the CDC's George Luber. Consider Phoenix, an urban oasis with a metropolitan area of 4.2 million residents, in the northern reaches of the Sonoran Desert and the capital of one of the last states admitted to the Union. Since World War II, Phoenix has mushroomed from a sleepy copper mining and cow town of less than 100,000 people into a concrete jungle that epitomizes urban sprawl with downtown skyscrapers and vast subdivisions, girded by highways, sweeping cloverleaf freeway interchanges, and a grid of streets spread over 500 square miles of flat land. With a hot, dry climate that shares much in common with Riyadh and Baghdad, temperatures there can rocket up to 120°F in the summer. In the days before air-conditioning and a construction boom enriched landowners who turned farmland into miles and miles of tract homes, the dry desert air would cool down to a more bearable 75°F at night. Families would unfurl cots on front lawns, douse themselves with citronella to ward off bugs and scorpions, and sleep under the stars. But no more: Thousands of acres of cement, asphalt, and blacktop have made Phoenix about 10°F to 15°F hotter than surrounding rural areas, and it now gets hotter earlier in the day, stays hot longer, and normally doesn't drop below 90°F at night in the summer.

"Phoenix is an island in the middle of a desert that's adapted very well through the use of mechanical cooling," Luber reflected. "But imagine what would happen if the power went out during a heat wave." It happened in

Chicago in 1995: The unprecedented demand for electricity led to brown-outs and blackouts, and at one point, 49,000 households had no electricity. And it happened again in July 2012, when the derecho downed power lines and left more than 3 million people without electricity—some for days—while temperatures hovered in the triple digits. "Where would people go?" Luber continued. "They can't escape by going east or west or even south to Tucson because they're surrounded by desert, and they won't even be able to pump gas. The only cool place is Flagstaff—but you have to get there first, and the interstate would be jammed. Suddenly, you're marooned on an island, and one that is roasting with a high percentage of elderly."

THE BURNING SEASON

When intense heat is accompanied by forest fires, the lethal combination can send death rates soaring because there is often no escape from the toxic smoke. In July 2010, Russia endured its most intense heat wave in 130 years of record keeping, with daily highs in Moscow passing 100°F, as compared to the normal summer average of 75°F. If that wasn't bad enough in a country that was unprepared to deal with such high temperatures, a severe and prolonged drought—the worst in a century—had dried out the underbrush in the surrounding countryside. The blazing temperatures were like lighter fluid on the desiccated kindling, sparking raging fires in the forests and underground marshes and peat bogs, which are collections of partially decayed vegetation with very high water content. Once ignited, these bogs become steamy conflagrations that can burn for years, belching out far more smoke and incinerating 10 times more vegetation than aboveground canopy fires. Peat bogs are extremely susceptible to combustion, and in the record heat, hundreds of fires spontaneously burst into flame every day in the region surrounding Moscow, where 12,000 bogs burned.

The fires razed 2,000 homes, ravaged more than 500,000 acres of land, and created an acrid plume of carbon-rich smoke nearly 2,000 miles long that smothered the capital in a poisonous smog for six straight days. Carbon monoxide levels jumped to nearly five times what is considered safe, while other airborne pollutants exceeded the maximum acceptable limits by nearly seven times. The combination of unprecedented heat and suffo-cating haze doubled the death rate to an average of 700 people a day in the city; the morgue, which could accommodate 1,500, was filled to near capacity, and more than 14,000 people died in July in Moscow alone. By the time temperatures dropped in mid-August, more than 56,000 Russians had lost their lives.

In a normal July in these northern latitudes, the sun shines for more than 18 hours a day, but the air was so thick with smoke that no sunlight broke through. Familiar landmarks like Moscow's Kremlin and Saint Basil's Cathedral were shrouded in a gray cloud for weeks, and dozens of flights were grounded or diverted from the city's airports because visibility was less than 300 feet. The tiny particles of smoke seeped into every corner of the city, even deep into the bowels of Moscow's poorly ventilated and noto-riously overcrowded subway system, which had been used as a bomb shelter during World War II.

I spoke with numerous Americans who were living in Moscow during the unprecedented heat wave, and they told me a siege mentality had gripped the entire city. In a teeming metropolis where few homes have air-conditioning, residents were sternly advised to stay indoors, tape up their windows, avoid physical activity, and wear surgical masks if they did ven-ture outside because breathing the foul air for three or four hours would damage the red cells in their blood as much as smoking two packs of ciga-rettes in the same amount of time would. The smell of burned embers hung heavily in the air. "Everywhere you went in the city, it was choked with smoke—even the Metro stations," recalled Erica Green, an American writer and mother of two young children whose husband was working at the US embassy. "We had air-conditioning and we shut the windows and drapes tightly, but it was still incredibly smoky inside our house. We all had terrible headaches and the kids were coughing, with upset stomachs and itchy eyes. We evacuated to Saint Petersburg because I was worried about my kids, and when we unpacked our suitcases, smoke came out."

Well-heeled Muscovites were somewhat insulated from the effects of the extreme heat, able to sequester themselves in air-conditioned homes, snap up cool luxury hotel rooms, or leave town—long-distance trains and pas-senger flights were quickly sold out. Many businesses shortened working hours or simply shuttered their doors, foreign companies evacuated staff, tourists cancelled their bookings, and some countries closed their consular offices. But for the millions of poorer residents, many of whom live in tiny, coffinlike apartments in bleak and shoddy Soviet-era concrete high-rises at the edges of the capital, there was no escape from the miserable, stifling heat. Even the simple act of opening a window to get some relief could prove fatal. When psychologist Tatiana Dyment couldn't get her 70-year-old mother on the phone, she cut short her vacation in Croatia and raced back to Moscow, only to discover her dead in the bathtub with cold water from the shower still spilling over her, according to the *Los Angeles Times*. The windows in the sixth-floor apartment were wide open, and the acrid smell of smoke permeated all the furniture. Doctors suspected the elderly woman had been dead for two or three days, attributing it to "acute heart insufficiency." But Dyment believed something else snuffed out her mother's life: "I think this smog from the forest fires killed her."

City officials were criticized for their abysmal incompetence in their response to the combined crises of heat and smog that fueled fatalities. Enraged critics charged that none of the normal measures to ease suffering and death during a heat wave—checking on the elderly and shut-ins who lived alone, opening cooling centers, or even increasing the number of ambulances available and equipping them with bottled water—were done. "The authorities had made zero preparations and people didn't know how to deal with it," recalled Marco North, an American graphic designer based in Moscow. "One news report said that wet bedsheets on the window would filter out the smoke and let the air in, which was completely idiotic. People were selling masks on the street for $20 each. They didn't keep out the carbon monoxide, but psychologically people felt like they were doing something. It was over 100 degrees every day for a couple of weeks, and you could barely see a block away if you were walking down the street. Luckily, my apartment had windows that closed tightly and I bought a couple of oscillating fans. Normally, they sell for about $10 apiece. I paid $200 for each one, and I had to fight people to buy them."

In the countryside throughout central Russia, the devastation was even worse. Entire villages were leveled by the fires, with more than 2,000 homes reduced to rubble and thousands left homeless. Some blazes moved so fast that people were engulfed before they could evacuate, while the police, military, emergency responders, and firefighters were stalled for hours on virtually impassable roads that had buckled in the hot weather. "All around us, everything was on fire," one septuagenarian who lived in a small town about 100 miles outside of Moscow told reporters. "Houses, trees, the ground itself—it was all in flames. Even the asphalt on the road seemed to be on fire. It was like descending into hell."

The consequences of the wild weather that gripped Russia rippled across the Middle East and Africa in a destabilizing domino effect that underscored the interconnectedness of the global ecosystems. The wildfires, driven by the sizzling temperatures and record drought, incinerated the Russian wheat crop, which reduced the harvest by 40 percent and prompted the Kremlin to ban wheat exports. The resulting shortage pushed world food prices to record highs, sparking unrest in Tunisia and riots across Africa, including in Mozambique and Egypt, the latter of which is heavily reliant on Russian grain.

The Russian heat wave was just one in a series of dangerous and bizarre weather patterns during the summer of 2010 that also saw torrential rains in China kill more than 1,000 and force 68,000 people to flee, two-pound hailstones pummel rural South Dakota, and once-in-a-century floods displace more than two million Pakistanis. Several elements played roles in creating the erratic weather—some were natural events and some were related to global warming, according to meteorologist Heidi Cullen, author of *The Weather of the Future: Heat Waves, Extreme Storms, and Other Scenes from a Climate-Changed Planet*. There was a "huge ridge of high pressure parked over Russia, which is consistent with a major heat wave," she said in an August 2010 interview with *Salon*. "And when there's a ridge like that, you've also got a trough somewhere, and . . . that somewhere is Pakistan, where all of that intense flooding is taking place." But the bad weather was intensified by an increasingly warm planet, Cullen said, and surface temperatures in the Indian Ocean that were about 2°F above normal, which is considered extremely warm.

Russia's leaders have long been notorious skeptics when it comes to climate change. They've purposely stood in the way of meaningful progress on curbing greenhouse gas emissions because they assumed they would be one of global warming's chief beneficiaries, scoring a more moderate climate, a longer growing season, and a thawed-out Siberia and Arctic, which would open up the Northwest Passage along the top of the globe and the region's vast stores of fossil fuels. But the disastrous summer of 2010 was a wake-up call with its devastating drought, fierce and relentless heat, and uncontrollable wildfires that scorched thousands of acres of forest and grain at a cost of $300 billion and suffocated half of European Russia in smog that claimed more lives than were lost in the 2003 heat wave. The haywire weather forced politicians in the oil-rich nation to finally admit the obvious: "Everyone is talking about climate change now," then-president Dmitri Medvedev, who wisely cancelled his holiday at a Black Sea resort when the heat wave hit, told the Russian Security Council. "Unfortunately, what is happening now in our central regions is evidence of this global climate change, because we have never in our history faced such weather conditions in the past."

Around the globe, fire seasons are already longer, the blazes are fiercer and hotter, and each year surpasses the previous one for record-setting fires. Here in the United States, 2011 was one of the worst wildfire seasons in history: 8.7 million acres of land burned, according to data collected by the federal National Interagency Fire Center—that's more than the areas of New Jersey and Connecticut combined. Texas was gripped by the most severe dry spell since George Washington was inaugurated president in 1789, according to tree ring data going back 425 years. In the Lone Star State, fires consumed 2.7 million acres—more than twice the size of the state of Delaware. But 2012 turned out to be even worse, with more than nine million acres incinerated as the Great Plains were gripped by an exceptional drought combined with high temperatures and strong winds. Even normally temperate states like Montana, Minnesota, and North and South Dakota experienced the driest September on record since 1895.

Experts say that changing weather patterns, rising temperatures, and increasing drought conditions have brought us to the threshold of a new era of monster fires, massive infernos that can rapidly spiral out of control. Three of the biggest blazes on record occurred in 2011. From May 29 through July 8, the Wallow Fire in eastern Arizona blackened 538,000 acres, making it the largest wildfire in Arizona's history, and forced the evacuations of nearly 10,000 people, shut down highways, and snarled traffic when fierce winds and parched weather caused the fire to swiftly spread. In Texas, 314,000 drought-stricken acres were scorched in the most destructive wild-fire in that state's history, and in Georgia's Okefenokee National Wildlife Refuge, more than 309,000 acres were incinerated.

These fires—which are made far worse by conditions resulting from global warming—also create an ominous feedback loop that amplifies heat-ing trends by pumping enormous amounts of carbon dioxide into the atmo-sphere, according to research conducted by scientists at the National Center for Atmospheric Research and the University of Colorado in Boulder. Using satellite observations of fires and computer modeling, they estimated that fires in the continental United States and Alaska release, on average, about 290 million metric tons of carbon dioxide annually, which is the equivalent of 4 to 6 percent of the nation's CO_2 emissions from burning fossil fuels for an entire year (though emissions can vary depending on the number and severity of fires). And especially large infernos can spew enormous pulses of carbon dioxide swiftly into the atmosphere. The "striking implication," the authors noted, is that a severe fire season that lasts only one or two months can be responsible for fouling the air with as much of this potent green-house gas as some states' vehicle emissions for the entire year. Consequently, more fires will beget even more conflagrations in a self-perpetuating cycle.

But it's not just drought-stricken vegetation coupled with warmer tem-peratures that generate these intense infernos. Throughout the Northern Hemisphere, pine forests are being decimated by bark beetles. In what is now the largest outbreak in recorded history—since 1997, bark beetles have killed six billion trees—these tiny predators, which are about the size of a grain of rice, have burrowed through the protective bark, drilled deep into the trunk, and infected them with a fatal fungus. The needles of the infected trees turn red, then gray, as the trees die, turning once-lush green moun-tainscapes into denuded wastelands that kindle raging forest fires.

In a 2013 study, US Forest Service scientists also found a connection between the loss of trees stemming from infestations with one type of bee-tle, the emerald ash borer, and increased mortality from cardiovascular and lower respiratory tract illnesses, which are the first and third most common causes of death in the United States. Researchers looked at 15 states that had experienced significant tree loss due to the beetle infestation and concluded the borer was linked to more than 6,000 deaths from respiratory problems and 15,000 heart disease–related deaths. There were "several plausible mechanisms linking these types of deaths with" tree loss, researchers noted, including poorer air quality, increased stress, and lack of physical activity.

The changing climate is largely responsible for the epidemic: The beetles are surviving the warmer winters and at higher altitudes and latitudes instead of dying off in frigid weather, while the longer summer seasons mean they can reproduce twice a year instead of once, according to Christine Wiedinmyer, an atmospheric chemist at the National Center for Atmospheric Research in Boulder. The incidence of these fungal infections is increasing exponentially, killing off hundreds of millions of pines on millions of acres of land as the scourge spreads across the West from New Mexico to the Yukon Territory in Canada. The beetles are even whittling through white pine forests in regions where they've never been seen before, like the Northern Rockies, particularly in Yellowstone National Park, and in the Western Cascades throughout Washington and Oregon. "In Canada, there are fears they could wipe out the whole boreal forest," said Wiedinmyer. "There would be no trees in Canada—one of our largest ecosystems and carbon sinks in the world—which would be really catastrophic. And there's nothing we can do about it."

THE PHILADELPHIA STORY

"Get Ready for a Scorcher," newspaper headlines all over eastern Pennsylvania warned as the region headed into a hot and muggy Memorial Day weekend in 2012. The thermometer was expected to soar to near 100°F, and the actual temperatures on paved surfaces that bake in the sunshine, it was said, might top 120°F. The National Weather Service issued an excessive heat warning for the holiday weekend—the equivalent of a code red alert. This prompted the City of Philadelphia to launch its comprehensive emergency heat plan and set in motion a series of events to prevent fatalities. Thousands of block captains and neighborhood buddies swung into action to check on elderly neighbors and shut-ins; outreach programs zeroed in on the homeless; hours at senior centers were extended; cooling centers were opened up throughout the city; eldercare facilities were required to maintain a temperate environment; the numbers of paramedics and other emergency personnel were beefed up; and utility shutoffs of customers in arrears were suspended to prevent the loss of electricity when people needed it the most. Newspapers and local news broadcasts featured regular reports throughout the weekend on how to ease heat exposure, and a city-affiliated Heatline was activated. In response to calls made before illness set in, nurses and health specialists were dispatched to the homes of vulnerable residents to evaluate their health and their ability to withstand the heat. "They may do something as simple as making sure a window that has been stuck closed is opened," said Jeff Moran of the Philadelphia Department of Public Health.

Philadelphia's pioneering program—now a widely copied model for responding to heat waves—was catalyzed by catastrophe: a 1993 heat wave where 10 days of sweltering weather claimed 118 lives. This was a much higher death toll than in other places that were baking under equally high temperatures, leading national broadcast outlets to call the city the "Heat Death Capital of the World." "That was unheard of because they were reporting a couple of orders of magnitude difference [in the numbers of dead] from other spots affected by the heat wave," recalled Gary Szatkowski, head meteorologist at the regional National Weather Service Forecast Office in Mount Holly, New Jersey. At that time, most coroners and medical examiners would classify a death as heat-related only if the core body temperature was above 105°F. But the gruesome reality is that some heatstroke victims aren't discovered for days—by which time body temperatures have cooled. Consequently, the Philadelphia Medical Examiner's Office decided that using just core temperature alone wasn't a good enough yardstick for calculating heat-related deaths and decided to include circumstantial evidence from the environment, such as whether there was any ventilation or air-conditioning in the room. The city's high numbers prompted a review by the CDC, which ultimately concluded the method was actually more accurate.

Philadelphia also put into practice an early heat warning system based on research by Laurence Kalkstein, who was then at the University of Delaware; civic officials believed these benchmarks were a better predictor of which weather conditions were more deadly. "The usual heat warning system is based on an arbitrary threshold of when the heat index hits 105 degrees," said Kalkstein, who is now at the University of Miami. "But there's no relation between that and the actual human response—people in Boston react very differently than residents in a more southern city, like Atlanta, yet the same standard was applied throughout the country."

Later adopted by the National Weather Service, these guidelines incorporated three crucial elements: a more nuanced view of how we respond physiologically to heat; the region's physical environment, for example, whether it's an urban heat island; and a complex classification system that pulls together a range of meteorological factors to identify the more-life-threatening weather conditions. Up to two days in advance, the system can spot oppressive systems composed of umbrellas of hot, humid air that migrate from tropical waters, giving local authorities precious time to launch lifesaving strategies. Kalkstein's team also observed that people were more vulnerable to heatstroke in cities that aren't accustomed to hotter weather, and that sharp spikes in temperature—a swing from, say, a tolerable 80°F to a run in the low 100s—can trigger more deaths. "Weather variability is a key factor, which is why more people die in Toronto than in Phoenix—people aren't used to intense heat waves that rarely occur, but when they do, they create havoc," said Kalkstein. "Different cities have different profiles, too, and lifestyle and architectural issues also contribute; the brick row homes in Philadelphia are less conducive to helping people through the heat than airier homes in the South."

Plus, when heat waves occur earlier in the summer, people are more likely to die because they're not yet acclimated to hotter weather. But as the summer goes on, our bodies gradually adapt by helping sweat glands produce more perspiration on the skin's surface, cooling the body. "By the end of summer, bodies have become more resilient," said Szatkowski of the National Weather Service. "The exact same weather conditions that might prompt us to issue a warning in early June wouldn't be considered as dangerous in late August."

Cities across North America, including Chicago, Toronto, Atlanta, Phoenix, and New York, have since adopted similar strategies to prevent fatalities. These programs seem to work: In 2010, only 15 people died in Philadelphia because of the heat, and other cities have experienced similar reductions in death tolls. The urban heat island effect can also be offset by increasing green space—rooftop gardens, public parks—and reducing black pavement. Chicago, to cite another example, launched a program to identify places in the city where heat tends to build up and figure out why, and then take steps to address the problem—introducing more green space with parks and trees, painting roofs so they reflect the sun, or planting gardens on them so they don't absorb heat. "We can't prevent the heat waves, but we do know how to stop the loss of life," said Luber. The real question is what will happen in the coming decades when places like Chicago have as many as 72 days over 90°F, making their climate more like that of New Orleans. Will these programs be enough, or will it become just too darned hot?

CHAPTER FOUR

HEALTH CARE
ON LIFE SUPPORT

It was sultry and hot when I traveled to New Orleans in June 2012, the start of the hurricane season, and air conditioners were going full blast across the city. Seven years after Katrina, the central core had been completely rebuilt; throngs of tourists were shopping in the narrow European-style streets of the French Quarter and filling up the jazz and blues clubs on Bourbon Street. In other parts of the city, however, the scars remained, with long stretches of abandoned and boarded-up homes and litter-strewn empty lots that will probably never be restored.

Yet when I visited the city's vast medical complex, located in the heart of the downtown business district, it appeared to be one giant construction zone. It is comprised of a handful of hospitals, including teaching facilities for the Tulane University and Louisiana State University medical schools, which are sprawled across several square blocks along a busy eight-lane thoroughfare, directly across from the gleaming, refurbished Superdome. Construction cranes dotted the skyline like giant birds of prey, and the incessant sounds of bulldozers and pile drivers filled the air as they laid foundations for the glass and steel high-rises. The complex will house two new state-of-the-art hospitals and anchor New Orleans's public health system of the future.

But tucked in a corner behind a barbed wire fence was Charity Hospital, which has remained vacant since Katrina. Like a faded grand dame, the soaring 300-foot-high gray building once towered over downtown both literally and metaphorically, but now it seemed greatly diminished, a ghostly remnant of the city's past. "Big Charity" was a New Orleans institution, one of the last of the fabled inner-city public hospitals that took care of all comers. Founded by a French shipbuilder, it opened its doors in 1736 when New Orleans was still a French colony—only New York City's Bellevue is older. When it was rebuilt in the late 1930s with three million of Franklin Delano Roosevelt's New Deal dollars, the sleek, art deco–style building was an architectural and technological marvel, and one of the largest and the most advanced facilities of its day. At one time, the teaching facility housed nearly 3,000 beds, making it the second-largest hospital in the United States, and trained 70 percent of Louisiana's physicians, including pioneering cardiac surgeon Michael DeBakey. The world-class Level 1 trauma center—among the busiest in the nation—drew young doctors from around the world, who learned how to treat stab wounds and gunshots in the crime-ridden city. It provided a sanctuary for the poor and disenfranchised for more than 250 years—until Hurricane Katrina made landfall along the Louisiana–Mississippi border on Monday, August 29, 2005.

Gathering steam from the overheated waters of the Gulf of Mexico, the Category 3 storm whipped up 140-mile-an-hour winds and created a 28-foot storm surge that steamrolled over coastal communities from Louisiana to the Florida panhandle. Hurricanes are a fact of life in New Orleans, and hospital staffers at Charity knew the drill—they camped out in the halls in sleeping bags and braced themselves for the hit. At first, the damage seemed relatively minor—parts of the hospital were flooded, windows were shattered by high winds, and some wings lost electrical power until the two backup generators kicked in. By Monday night, the storm had passed and there was a palpable sense of relief that a bullet had once again been dodged. The plucky hurricane activation crew that had volunteered to staff the hospital during such emergencies figured they would be heading home soon.

We all know this story doesn't end well, and what happened next has been meticulously documented in hundreds of media accounts. But there has been little examination of the lingering physical and psychological trauma in the storm's aftermath over the ensuing weeks, months, and even years after New Orleans was leveled in one of the worst natural disasters in our nation's history. What followed was an unprecedented collapse of the public health system in a modern American city—even as many civic leaders and health professionals acted with heroic intentions and almost unlimited resources were poured into rebuilding. Seven years later, when I spent time with many of the shell-shocked survivors of this disaster, it was clear the city still hasn't recovered.

In the early morning hours of Tuesday, August 30, as has been widely recounted, several of the levees and floodwalls of the complex 350-mile system of barriers that protected the largely below-sea-level city from the surrounding Mississippi River and Lake Pontchartrain failed under the extraordinary strain of Katrina's winds and rains, flooding the city with 15 feet of water. And the power outages that followed the flood brought to a halt many of the pumping stations that would have removed some of that water. When Dr. Ben deBoisblanc, head of the intensive care unit on Charity Hospital's 12th floor, looked out the window shortly after daybreak, he saw the entire city was underwater. Hundreds of terrified people were marooned on rooftops after using axes and hatchets to escape through attics when the rising waters engulfed them and the pockets of air ran out. "That's when I knew we were in big trouble," he told me.

Floodwaters swamped the hospital's basement, submerging the electrical generators and threatening the lives of patients in the emergency room on the first floor, which was taking on water. Suddenly, more than 1,300 people, including patients, doctors, nurses, support staff, their families, and people from the community who had sought shelter, were stranded with no power or water pressure. Toilets overflowed with waste, creating an overwhelming stench in the insufferable heat. The basement morgue was flooded, so patients who died were left in their beds or moved to stairwells and hallways. Cell phone networks and hospital phone lines went dead, cutting off their connection to the outside world.

With no electricity to power their equipment or elevators, the hospital staff carried more than 50 critically ill patients up a darkened stairwell where temperatures were pushing 100°F to a refuge in a second-floor auditorium. For the next four days and nights, medical personnel worked in battlefield conditions without the use of basic critical care equipment and miraculously kept alive desperately ill patients who had been on mechanical ventilators by squeezing air into their lungs with manual resuscitation bags, sometimes in total darkness. Because it was so hot, emergency room staffers slept on the 13th floor, though they barely got more than four or five hours' rest. "But by the time you hiked up there, you were soaked and there was no running water to take showers," recalled Dr. Peter DeBlieux, who ran the ER and taught emergency medicine to young residents. "The last three nights, we slept on the rooftop."

For three days, Charity Hospital, one of the biggest facilities in the country, had somehow fallen off the government's disaster management radar screen, and no help from the Federal Emergency Management Agency (FEMA) or the National Guard arrived. "On day three and day four, we were told, at least twice, 'Pack your bags, they're coming to get us,'" DeBlieux recalled. "And then they didn't. We heard on the portable radio that we had already been evacuated. That was priceless."

In the storm's chaotic aftermath, power lines fell, cell phone towers collapsed, and the communications infrastructure was decimated, depriving the city of a dependable network for coordinating emergency response operations. Health care and emergency services providers faced daunting challenges getting gravely ill people the care they needed. A makeshift triage center and temporary morgue were set up at the Louis Armstrong New Orleans International Airport west of the city, where patients were stabilized and then airlifted to hospitals throughout the South. Commercial parking lots and vacant stores located on high ground became mobile treatment centers. Several Navy hospital ships were dispatched to the New Orleans harbor.

But it was a herculean task. Most fire and police stations were underwater, dispatchers were unable to relay messages, and police and firefighters labored mightily to do their jobs under unimaginably adverse conditions, often without adequate equipment or trucks, food or water, a change of clothing, or a place to sleep. The entire New Orleans Fire Department established a base camp on the four-acre campus of a small Catholic college in Algiers across the bridge, on the west bank of the Mississippi, that had escaped flooding, and firefighters remained there for the next six weeks. Fire companies all over the city were told where to report by runners on foot, since this was the only reliable means of communication.

On Tuesday, a team of firefighters plowed through floodwaters to the Saint Bernard Projects to rescue people still stranded in one of the city's poorest neighborhoods. "It looked like a Third World country—people were floating by clinging to wood planks or ice chests," recalled Billy Shanks, a fire department captain. They began the arduous process of airlifting the old and the infirm on choppers that could land on a nearby bridge. "This elderly blind woman with diabetes had her little 10-year-old granddaughter with her, who was obviously [serving as] her eyes," said Shanks. "When we put her on the helicopter, the poor woman started crying for her grand-daughter so we put the little girl in her arms. That broke my heart." Later that night, driving down Magazine Street, a normally busy boulevard lined with antique stores, art galleries, and upscale boutiques, "it looked like the world had ended," Shanks recalled. "It was pitch black and there wasn't a light on in the entire city."

Looting, violence, and lawlessness spread throughout the city while fran-tic residents pillaged grocery stores for food, supplies, and water. Sniper fire directed at rescue helicopters, relief workers, and police officers interfered with evacuation efforts. "I was never afraid of wind, water, fire, hunger, or disease," noted Dr. Ruth Berggren, an infectious disease specialist at Char-ity Hospital who chose to stay with 18 gravely ill patients on the ninth floor of the hospital during Katrina. "My moments of fear came when I was con-fronted by agitated, fearful human beings bearing firearms. My husband"— an oncologist at nearby Tulane University Hospital—"was exposed to sniper fire twice while helping to evacuate the emergency room dock. People with guns shut down an entire hospital evacuation for many hours."

Conditions were equally grim at the 10 other flood-isolated hospitals, which collectively housed 1,800 patients and more than 7,600 people, including heath care professionals, their families, and area residents. Memorial Medical Center, spread over three city blocks, is a sprawling nine-building complex linked by elevated, enclosed glass walkways. It is located three feet below sea level in an older neighborhood near the river, about three miles from the French Quarter. More than 2,200 people were trapped inside the 350-bed facility by eight feet of floodwaters, including 200 patients, many of them seriously ill, and hundreds of pets belonging to hos-pital staff and people seeking refuge. By Wednesday morning, they had lost all power and the situation inside the hospital rapidly deteriorated. A chapel on the second floor, pressed into service as an overflow morgue, started filling up with corpses.

But there simply weren't enough boats and helicopters to rescue everyone right away because emergency services were overwhelmed by the demand and the evacuation process was agonizingly slow. Ultimately, 45 of the patients died, but Dr. Richard Deichmann, the hospital's chief of medicine, believes at least some of the blame for these deaths can be attributed to a series of bad decisions by local, state, and federal officials. First, Louisiana state troopers limited the hours of evacuation from the hospital by boat because of the looters roaming the neighborhood. Then, federal authorities halted evacuations by a private company—the ultraefficient Acadian air ambulance service that was doing a splendid job by all accounts—while law enforcement officials continually rebuffed offers of help by private citizens with boats. "Nobody ever envisioned that the authorities would prevent people from coming in to rescue us," Deichmann said during a recent interview in his office at Ochsner Medical Center, where he is now an associate director in the primary care department.

Despite having been repeatedly assured by government agencies that they would be evacuated, on Thursday morning the remaining staffers at Memorial were told by local authorities they were now on their own. "It was a phenomenal blow to hear that nobody was coming to get us," recalled Deichmann, who felt a deep sense of abandonment and betrayal. Fortunately, by Thursday afternoon, Tenet Healthcare Corporation, the hospital's parent company, dispatched a fleet of privately owned helicopters to ferry the last survivors to safety—a step they said they would have taken days sooner if they had been alerted by civic authorities. Earlier in the day, Coast Guard helicopters had arrived and volunteered to spend an hour transporting healthy people to safety (they didn't have enough space for seriously ill patients on stretchers). "Part of the disaster was natural, but a big, big part of it was man-made and poor decision making," said Deichmann. "We had plenty of resources to help people out for two or three days. But when it went on and on like that with no end in sight, that's when things became a lot more serious."

Inside the Superdome, the massive 27-story stadium downtown that sheltered 30,000 refugees for six horrific days, it was utter anarchy. The sweltering, dusky air was steeped in both the sweat of exhausted, unwashed people who were starving, frightened, and dehydrated and the stench of the overflowing toilets and huge piles of rotting trash baking outside in the sun. Food, water, and medicine were in scant supply, and some of those who sought sanctuary from the four-foot-high waters surrounding the 77,000-seat arena had arrived with just the shirts on their backs. Once inside the Superdome, people were trapped by armed guards. "It was like a war zone and very disorganized—the conditions were like something I'd seen in a remote region of Africa," said Dr. Seyed Hosseini. "I couldn't believe this was America."

The physician and his girlfriend, Lynda Wilson, had both been in graduate school in New Orleans—Wilson had just received her PhD at LSU while Hosseini was completing a pathology residency at Tulane after years as a general surgeon in his native Italy. Warnings about the severity of the coming storm prompted them to leave in what was the first mandatory evacuation in the city's 300-year history. "The mayor came on TV around noon and said 20,000 body bags were being sent to New Orleans and if you remain in the city, you needed to write your name and your Social Security number on your body," recalled Wilson. "We left late Saturday afternoon but we didn't take much except our dog—we had no clue we'd lose everything."

They joined the more than 1.3 million people fleeing the city and much of south Louisiana in cars, rental vans, trucks, boats, buses, and trains, all crammed with people, pets, and valuables in the largest dislocation of American citizens since the Civil War. It took the couple nearly eight hours driving at a crawl on the virtually gridlocked interstate to make the trek to Baton Rouge, which is about 80 miles upriver from New Orleans. The sleepy state capital doubled in size overnight with the arrival of 200,000 newcomers who booked every available hotel room, crowded into makeshift shelters, and camped out in the garages, hallways, and kitchens of extended family members, friends, or the legions of Good Samaritans, like the local dentist and his family who took in Lynda and Seyed.

When Hosseini heard over the radio that medical professionals were needed back in New Orleans, he contacted FEMA, which transported him back to the Superdome the next morning. But he quickly discovered that he was one of only two doctors in a stadium filled with thousands of famished, thirsty, traumatized people roasting in 100°F heat. He barely knew where to begin. "At least 1,000 people were in terrible shape with serious health care problems—uncontrolled diabetes, heart disease, high blood pressure, elderly people in wheelchairs with no medication," he recalled. "But people were scared that if they went out on the street they would be shot." Hosseini did his best to treat them with the scant supplies on hand, but he remains convinced that "people died because of the negligence. There was truly nothing—no food, the arena had no lights, and you couldn't even use the restroom."

At daybreak on Thursday, some FEMA workers at the Superdome hatched a secret escape plan and invited Hosseini to join them. Exhausted from lack of sleep and deeply discouraged by the deplorable, inhumane conditions, he decided to accept their offer, especially since two Tulane doctors had arrived the night before to relieve him. "They gave me a stick and told me to hit people if they tried to get into the truck because they were so desperate to get out of there," he recalled, still horrified at the memory. "I remember thinking, 'I'm a doctor. I don't hit people with sticks.'" They drove up to Baton Rouge and Hosseini spent the next four weeks volunteering at the Louisiana State University campus, where an athletic arena had been turned into the largest triage center and acute-care field hospital in US history. Under the direction of FEMA, the 800-bed facility was staffed mostly by volunteers, including about 1,700 health care professionals from across the country, who treated more than 6,000 patients and triaged more than 15,000 who had arrived by helicopter and ambulance from New Orleans.

On Wednesday night, doctors at Charity were finally able to begin evacuating the sickest patients to the airport, a process that continued until Friday afternoon. The pace of evacuation from here and other facilities engulfed by floodwaters was exceedingly slow because only boats and helicopters could be used, and they often could take only one or two patients at a time, with each round trip sometimes taking an hour or more. Patients had brief medical records taped to their forearms, and then they were transported by canoe or high-water trucks through the floodwaters to the rooftop of the parking garage of nearby Tulane Hospital, where residents, nurses, and respiratory therapists kept 50 critically ill patients alive until they could be airlifted out.

Patients were triaged and color-coded according to their potential for survival: black meant moribund—to be given comfort care only; red meant critical care was needed, while the greens and yellows indicated those who were most likely to survive. Doctors operating under extraordinary pressures made controversial life-and-death decisions. DeBlieux found out months later that many of the patients he had thought could be saved had been reclassified as black once they arrived at the airport. "We were devastated," he recalled. "The physician that I had passed the patients off to at the airport came up to me at a conference and burst into tears. 'I'm so sorry, I'm so sorry,' she told me over and over. 'We let you down.'"

After the Storm

New Orleans civic leaders, in tandem with federal agencies, had run an extensive disaster drill less than a year before Katrina hit, but it did little to prepare them for the devastation they confronted once the waters began to recede. The damage to the public health system was staggering: Only 3 of 16 acute-care hospitals were able to reopen. "Big Charity" and University Hospital were shuttered—Charity remains closed to this day—and more than 70 percent of physicians' practices citywide were forced to relocate. On the Sunday morning after the storm hit, Mike Leavitt, the US Secretary of Health and Human Services, held a meeting with the leaders of the three open hospitals on the tarmac at the New Orleans airport to talk about restoring acute-care and emergency services; it was the first of what became daily morning conferences with the leaders of active hospitals, emergency medical services, physicians, and the military.

But they were clearly flying by the seat of their pants. "We didn't actually have a playbook, and there was really no one in charge, so that made everything a lot more chaotic and confusing," recalled Dr. Karen DeSalvo, who at that time was an administrative dean and head of the department of internal medicine at Tulane Medical Center. "At the end of one meeting, this military guy from the 82nd Airborne"—the famed disaster relief unit—"came over and told me, 'We need to open Tulane Hospital because we don't have any facilities downtown, and I've been working on a checklist for what we need to do to open a hospital.' He opened up his little notebook and he's got a handwritten checklist. I had this horrible sinking feeling that if the military didn't know how to open a hospital and they were making it up on the fly in little notebooks, we were in a lot of trouble." She continued, "Now there *is* a list for how to open a shuttered hospital that's been codified, but at that point, we were making it up as we went along."

For months, health care remained "unacceptably primitive," recalled Berggren, with serious shortages of hospital beds, laboratory facilities, equipment, supplies, and crucial staff, from nurses to specialists like anesthesiologists. Tulane and LSU lost up to a third of their medical school faculties, who were either laid off or simply moved elsewhere because they had no assurances of jobs. Doctors moved to storefronts and military tents and stored medications in ice chests. Three members of the US Congress warned of a "potentially catastrophic public health crisis" because the city's medical care infrastructure was so precarious and stretched so thin that it could be overpowered by a single bus crash or an influenza outbreak.

At that time, Hurricane Katrina was the most costly natural disaster in our nation's history, totaling more than $81 billion in damages and up to $250 billion in economic losses. The storm also exacted a crushing human toll: More than 1,800 people died as a direct result, most of them the homebound elderly who were trapped by the floodwaters. More than a quarter of a million people were left homeless, and many of them, stranded for days without food, water, shelter, or access to medical care, barely survived the storm's harrowing aftermath. About half of them were permanently uprooted, even though many came from families that had lived in New Orleans for generations. They resettled in Sunbelt cities like Houston, Dallas, and Atlanta that were hundreds of miles—what might as well be light-years—away from the vibrant French-flavored cultural gumbo of the Big Easy, which thinks of itself as the northernmost Caribbean city.

Medical records at many hospitals had been kept in manila folders rather than electronically, so health care providers typically had little information about what medications displaced patients with life-threatening ailments like cancer or heart disease were taking, or even their medical histories. The losses were felt most keenly by the neediest: When Charity was abandoned, they lost a major safety-net hospital. Many died from dehydration, stress, or simply from lack of regular medications. Chronic ills like high blood pressure, AIDS, and diabetes went untreated, leading to potentially fatal complications like diabetic ketoacidosis. Kidney dialysis centers that had been serving more than 3,000 patients were leveled. There were lingering effects from the growth of toxic molds, which greatly exacerbate asthma, allergies, and other respiratory ills, especially among children. The contaminated floodwaters—a soupy stew of chemicals from cars, machinery, gas stations, pesticides, solvents, and household products— contributed to skin infections, respiratory problems, and persistent staph infections. Exposure to this poisonous brew can also increase the risk of cancer later in life. City mortality increased significantly, with deaths spiking by as much as 25 percent over the previous year, even though one out of four residents had fled the city. Because so many had lost their jobs, private hospitals hemorrhaged millions in caring for uninsured or indigent patients without compensation.

Felice Guimont had been an ER nurse at Charity, and like many other educated professionals, she lost her health insurance when the city's economy collapsed. A tall, expressive woman with long dark hair and incandescent blue eyes, she's had type 1 diabetes since childhood, requiring up to five shots of insulin a day to manage her illness. But she has struggled since the storm because she can't find full-time work and has had only spotty insurance coverage. "Everyone thinks nurses are rich, but the middle class in New Orleans got wiped out because we didn't know how to work the system—we never needed to before," said Guimont, who's the only one in her large, extended family that didn't lose everything in Katrina. Her aged parents had raised eight children in their home of more than 50 years in the Lower Ninth Ward before floodwater inundating it forced them to flee to her sister's house in Tennessee. Guimont stayed with another sister in Beaumont, Texas, where doctors and drug company reps were dispensing insulin to Katrina refugees at a local shelter. But others weren't so lucky, Guimont said: "A lot of people died because they didn't have the right medication."

Others with chronic ailments like tuberculosis failed to get adequate care, and by the time they returned six months later, their illnesses had come roaring back. "The best way to get drug-resistant TB is to interrupt therapy," said Berggren, the infectious disease specialist. "With all these displaced people, we were very fortunate there wasn't a major TB outbreak."

Even nine months later, Berggren couldn't order local lab testing when she suspected someone had tuberculosis. The sputum samples had to be processed in a Texas lab nearly 500 miles away, or she could put the patient on a bus to Baton Rouge and risk exposing the other passengers to it. "The city couldn't get it together to make TB smear samples available to poor people," she recalled. "In the meantime, the guy with TB is coughing in the front of my office."

A year after Katrina, New Orleans's population had plummeted by half. Crime, suicide, and depression rates were doubled. Alcohol abuse and domestic violence jumped. Post-traumatic stress disorder was rampant, with about 10 percent of the population crippled by debilitating mental anguish. Many had lost loved ones, their homes, their jobs, and all their worldly possessions and were separated from their families. A few had undergone horrific ordeals, such as having to be rescued from roofs or bridges or witnessing dead bodies floating in poisonous waters. Some were living in emergency shelters, which forced another layer of hardship on them, or in toxic FEMA trailers in fractured communities that lacked the most basic amenities—public transportation, grocery stores, schools, churches, parks, even gas stations—that create the normal fabric of a community.

Fire department captain Billy Shanks's wife, Flora, left New Orleans with their two young children when Katrina was still gathering steam in the Gulf and stayed with friends in a Houston suburb for more than three months. They lost everything when the 17th Street Canal along Lake Pontchartrain breached, covering their close-knit neighborhood of brick bungalows, wood-shingled cottages, and tidy lawns with 10 feet of oil-slicked water carrying sewage and rotting dead animals. They had no inkling then that it would be two years before their flooded house would be rebuilt and they could move back in. Flora called Billy shortly after they settled in Texas: Their kids' classmates at their new school had taken up a collection and their parents had bought the family Wal-Mart gift cards. "It reminded me there is a world out there and people actually care," said Shanks, who still chokes up at the memory. He stopped by a shelter shortly after the flood and got a change of clothes, a present for his wife, and a couple of toys for his kids. "All I had was the uniform I had on," he said. "I'd drive to Houston every couple of weeks and I'd think, 'I'm homeless.' My pride and joy was my house, and it was now gone."

The profound sense of dislocation Felice Guimont and the Shanks family experienced was widespread. A city that reveled in an image of *bon temps* now seemed shrouded in an inky veil of gloom. "People were very depressed—at least half of my patients were on antidepressants," said Deichmann. "You felt very isolated, and a lot of friends left who didn't come back, so social networks were disrupted. Just getting the easiest things done that we take for granted—going to the grocery store, getting gas for your car— was a huge deal. My house wasn't flooded, but it took a year to get the mail delivered again—and services got cut off because we didn't receive our bills. It was a horrible situation."

Yet the mental health care infrastructure—which had been inadequate before—was virtually nonexistent at a time when the need couldn't possibly have been greater. At one point there were only 22 psychiatrists in a city of 200,000. Within a year after Katrina, five doctors became so despondent they took their own lives. "It wasn't just the destitute poor who had no hope, but professional people who didn't leave New Orleans and who stayed in the middle of it," recalled Berggren. "One woman who was a neighbor of ours was a pediatrician and a mother of three beautiful girls. In the aftermath, she was practicing in a temporary trailer arrangement and just saw so much despair that she fell into a clinical depression and committed suicide."

Dr. Ben deBoisblanc, a courtly 50-something southerner with sharp, almost birdlike features and tousled blond hair, spent the two months following the storm sleeping in his SUV while working shifts at a hospital in Baton Rouge. "I wanted to come back and start rebuilding, but there was nothing—no gas stations, no grocery stores—none of the services that make everyday life possible," he recalled. "Life as you understood it just ground to a halt."

"The hardest thing I've ever done in my life was not Hurricane Katrina," he said during an interview in his spacious office at Louisiana State University's medical school complex. "The hurricane itself was a gift to me, an opportunity to do something really special and to watch people acting their best, doing heroic things that were really moving. But the next year and a half was just awful—it was like watching the grass grow because nothing happened.

"We had a meeting of the medical center staff about two weeks after Katrina—everyone who had been at Charity Hospital, at LSU, at Tulane Hospital, it was like six or eight hundred people," he continued. "It was as though you had the first reunion of your army platoon after a firefight and you see your old buddies again. The acting chancellor gives this speech that we're going to rebuild better than ever and he got everybody fired up. And then nothing happened. It was February or March before we could even get back into this building. The neighborhood I lived in was black for six months—no lights, no electricity, nothing. I reconnected with some of my friends and we had an informal group that would meet by the lake to run in the morning and we called it the 5:20 Club. Well, at 5:20 in the morning it was pitch black, surreal. We could have been in the middle of Afghanistan. There were packs of wild dogs and feral pigs and you're thinking, *This used to be a city*. It was like one of those Armageddon movies where you're the lone survivor."

For nearly a year, ER doctor Peter DeBlieux practiced medicine in a 30-bed military tent that could accommodate about 150 patients a day, which is roughly a third of the number that came in every day to Charity. He and other ER staffers wandered the city like gypsies, setting up shop first in the parking lot of University Hospital, right next to where Charity stood empty, then at the city's convention center, and finally in a makeshift emergency department and walk-in clinic in an abandoned Lord and Taylor store. "I was livid and it's psychologically debilitating," said DeBlieux. "A lot of doctors were displaced who didn't come back—and who can blame them? They'd be coming back to a broken infrastructure for themselves and their families, so there's no incentive. Because the volumes were so great here on physicians, no one was offering second opinions. Up until quite recently, we were encouraging friends and families with new diagnoses of complicated cancers to seek care elsewhere. For a while we had only one Level 1 trauma center. If you're in a car collision or have a heart attack or a stroke—which are all time-sensitive conditions—and don't get definitive care within the hour, your outcomes are poor. And if you're stuck waiting for 10 hours in an emergency room instead of 3 hours, that's bad, too. This was a great leveler because it didn't matter if you were rolling in cash and had plenty of insurance—health care here was limited and there's no question that it increased mortality."

CAN IT HAPPEN AGAIN?

Much has been written about the colossal failures of FEMA in the wake of Hurricane Katrina. It would be comforting to think that what happened in New Orleans was an isolated episode, the result of a perfect storm of gross incompetence up and down the line on the federal, state, and local levels, that wouldn't be repeated should a crisis of this enormity occur again. Clearly, there were catastrophic breakdowns in communications networks, fumbles early on by government officials that fueled perceptions of callous indifference and bumbling ineptitude, and bureaucratic screwups that stymied rescue efforts and caused needless suffering and deaths. Indeed, a US Senate investigative report issued in April 2006 concluded that FEMA was a complete shambles and the deeply rooted failings and waste at every level of the agency were "too substantial to mend."

But the reality is that the Hurricane Katrina rescue effort, the largest and fastest response of its kind in American history, is widely considered "an extraordinary success—especially given the huge area devastated by the storm," according to an investigation done by *Popular Mechanics* magazine. To put this into perspective, previous computer simulations had predicted that a Katrina-strength hurricane making landfall near New Orleans could claim as many as 30,000 lives. But the official death toll was about 1,800 people in the entire state.

Within 72 hours of the storm's landfall, more than 100,000 emergency personnel arrived on the Gulf Coast. A flotilla of military and police choppers filled the skies, adroitly plucking hundreds of people off rooftops and balconies and ferrying perilously ill patients to safety. Firefighters and National Guard troops cleared away debris to make hundreds of miles of streets and roadways passable. FEMA, working in tandem with the Louisiana Department of Wildlife and Fisheries, the state police, and local sheriffs' departments, dispatched an armada of water rescue teams. Within a week, 50,000 National Guard members rescued 17,000 people, 4,000 Coast Guard staffers saved 33,000, and firefighters rescued another 17,000, while the boat-borne crews helped another 20,000 people to safety.

From that perspective, what happened in New Orleans and the surrounding Gulf Coast communities is first and foremost a cautionary tale about the collapse of the public health system in the wake of a natural catastrophe. Despite the heroic efforts of health care professionals and the massive evacuation and rescue operations, many people simply refused to leave, the mostly indigent population that didn't have the resources to go somewhere else were left in harm's way, and there wasn't an adequate plan for making sure hospital and nursing home patients were transported to safety. In addition, much of the medical infrastructure—hospitals, clinics, doctors' offices, laboratories, and other support facilities—was damaged, some of it beyond repair, and it was years before it was rebuilt. Given the similarities in the systems from city to city, just imagine the deleterious impact to the health care infrastructure if a calamity of the same magnitude struck Houston, Atlanta, Miami, or another large population center in the hurricane belt, or even a city vulnerable to storm surges during heavy rains, like Los Angeles or San Diego.

In fact, one doesn't have to imagine it. What happened in New York City when Superstorm Sandy hit the East Coast in October of 2012 is a case in point. Health care officials had had seven years since Katrina to make preparations based on the lessons that should have been learned from the earlier calamity. But events there eerily echoed what happened in New Orleans and revealed the Achilles' heel in many of the best-laid disaster plans, especially at facilities in low-lying flood-prone areas. Backup generators failed at one of the nation's premier hospitals, New York University Langone Medical Center, a sprawling complex in mid-Manhattan near the East River, forcing the evacuation of all 215 patients to nearby facilities in the midst of gusting winds shortly after Sandy made landfall on Monday, October 29. Several blocks away, at Bellevue, the nation's oldest public hospital, the basement flooded and the hospital had to resort to backup power. By Tuesday, all 32 elevators at Bellevue had stopped working because of flooding, and the National Guard was called in to help evacuate their 500 patients—a process that took nearly two days to complete. Receiving hospitals, like Mount Sinai Medical Center in northern Manhattan and Brooklyn's Maimonides Medical Center, strained to care for the overflow and to treat deathly ill patients who arrived with scanty medical records.

Once the storm subsided, thousands of public health nurses, who themselves had trouble navigating in the city because of a lack of public transit, power outages, and heavy traffic, were deployed throughout the city. But they struggled mightily to meet the needs of tens of thousands of chronically ill patients dependent on home medical care who were stranded in high-rises with no heat, electricity, phone, or elevator services. Going without proper medication can be catastrophic for frail people with diabetes, asthma, high blood pressure, or other silent killers. Many of them probably did and will experience the same deprivations, challenges, and heightened rates of serious complications and mortality that residents of New Orleans did for years afterward.

Much of the city slowly shuddered back to some semblance of normalcy in the days following the storm, especially in Manhattan where power was restored to the lower half of the island within the week. But in some of the hardest-hit communities, densely populated places like Far Rockaway and Breezy Point, hundreds of homes were shattered, public housing projects were left without running water or power, doctors were flooded out, and pharmacies were closed. As a result, scores of sickly residents were left in perilous straits, cut off from access to medical care. In response, for the first time in its fabled 40-year history, Doctors Without Borders, the Nobel Peace Prize–winning charitable organization that parachutes volunteer health care professionals into disaster zones like Haiti, Rwanda, and Indonesia, set up a clinic on American soil, in Far Rockaway, to care for all the people who had slipped through the cracks of the governmental safety net.

Even two months later, hundreds of elderly and disabled New Yorkers who had been hastily evacuated from nursing homes or assisted living facilities in seaside communities were still sleeping on cots in makeshift quarters and remained without regular changes of clothes at hotels, halfway houses, and overcrowded rehab, adult care, and skilled nursing facilities all over the city. Many of these places were swollen to double their capacity to accommodate the temporary residents. "It feels like a MASH unit here right now," one staff member told the Associated Press just before Christmas. "People are working incredibly hard. The circumstance could not be more dire, and people are getting the best possible care we can manage."

While New York City's overall emergency response to Sandy was impressive, a disaster of this magnitude illuminates the fissures and critical deficiencies in the public health system, especially for our most vulnerable citizens. But what happens when a hurricane the size of Katrina or Sandy—or worse—isn't just a once-in-a-decade or even an annual event? When Category 5 storms rampage across the hurricane belt with a ferocious frequency that pushes our public health system beyond its limits? It's not that far-fetched, according to growing scientific evidence.

Climatologists have long said they can't blame a specific weather-related disaster, such as Katrina, on climate change, even though an increase in the frequency of these extreme weather events as the planet heats up is "consistent" with their computer models. But more recently, there's been a dramatic shift in scientific assertions on this controversial topic. The string of severe disasters around the globe cannot be explained by natural climate variation, recent research conducted by the UK's Hadley Centre for Climate Prediction and Research showed with a high degree of confidence: The greenhouse gases dumped into the atmosphere by our modern industrial society have set us up for a new normal of even more torrential rains and more severe storms.

Already we know Earth's temperature has increased about 1°C (1.8°F) since the Industrial Revolution. Sea levels—which have swelled by about seven inches since 1900—will continue to rise as the planet heats up, a change attributable to a couple of factors. First, glaciers are melting at unprecedented rates, pouring more water into the oceans. (In September 2012, Arctic ice shriveled to 1.32 million square miles, the smallest extent ever recorded and less than half the area it occupied four decades ago.) In addition, as the oceans warm the water expands, which has pushed up sea levels by another six inches. Moreover, because warm air is lower in density than cool air, it rises like a balloon. On humid days, the hot air carries with it buckets of water vapor. As temperatures continue to climb, the amount of moisture the atmosphere can contain goes up, too—about 7 percent for roughly every degree Celsius.

While more moisture inevitably means more rain, familiar precipitation patterns will change, according to computer models, yielding fiercer, drenching downpours rather than gentle summer sprinkles. It is basic atmospheric physics: Rain occurs when the moisture in the atmosphere cools off enough to condense water vapor into a liquid. But high levels of greenhouse gases have collected in the outer layers of the atmosphere, called the troposphere, blocking the normal release of radiation into space, which makes the atmosphere's normal cooling mechanisms less efficient. Consequently, according to projections from the Hadley Centre, it will rain less overall, but when precipitation—whether snow or rain—does happen, it will be much more intense.

Heightened moisture and energy in the atmosphere combined with increasingly warm ocean temperatures translate into more powerful hurricanes. The way in which Katrina exploded from a modest tropical storm into a whirling dervish of devastating fury is a good example. Temperatures in the Gulf were 2°F to 4°F higher than normal because of the lack of cloud cover that normally prevents the sun from heating up the waters—this was a significant increase in energy that turbocharged the storm. And these conditions were the result of a high-pressure system known as the Bermuda High, which is usually the culprit behind the hot, muggy weather in the Mid-Atlantic part of the country. But this regularly occurring weather system has now expanded to the south and west, which fuels the storms, like Katrina, that now lash the Gulf's northern coast.

Ocean water must remain above 82°F for hurricanes to form. As ocean temperatures continue to rise past this threshold, the hurricane belt will spread and intensify, sparking powerful storms farther up the Atlantic Coast that threaten even New England. Superstorm Sandy, for instance, was caused by a hurricane being driven by higher ocean temperatures crashing into an exceptional cold front triggered by an unprecedented thaw of vast sheets of Arctic ice. More than half of the US population lives in coastal cities—about 153 million people, according to the National Oceanic and Atmospheric Administration, and that figure is expected to reach 165 million by 2015. The majority live in regions vulnerable to storms like Katrina and Sandy, along the Atlantic and Gulf coasts, a report by the National Academy of Sciences reveals, which greatly increases the potential of hurricanes to cause "catastrophic loss of life and property damage."

But it's not just the rising sea levels and increasingly intense storms themselves that pose a threat. Government flood officials believe our current flood protection system is woefully inadequate to safeguard homes, farms, industrial zones, chemical plants, and refineries from the coming storms. The patchwork system of levees and walls that collapsed after Katrina were, according to a report by the US Army Corps of Engineers, "a system in name only." What happened in New Orleans could easily occur somewhere else. Right now, there are 30,000 miles of levees in the United States— enough to girdle the globe. But most of them, nearly 70 percent, are not engineered well enough to hold back serious flooding, according to a 2010 FEMA report, and very few could withstand the more severe storms and heavier floods that climate scientists anticipate. In fact, FEMA predicts that rising temperatures will increase the total size of floodplains nationwide by about 45 percent, a staggering rise that places millions of Americans in jeopardy.

Overall, the price tag for Katrina's massive rescue and later recovery efforts exceeded $200 billion, which doesn't include the $80 billion in damages left in the storm's wake. The bill for Sandy's devastation is already well over $65 billion, and the tab for reconstruction should easily surpass what was spent in New Orleans. Since 1980, in fact, the frequency of multibillion-dollar weather disasters has at least doubled, according to the US National Climatic Data Center in Asheville, North Carolina. What would happen if there was nowhere to run? What if we had no choice but to stand our ground when 200-mile-an-hour headwinds and driving rain from Category 5 storms come barreling in our direction? What if there was nothing left in the governmental kitty to pay platoons of military troops, firefighters, police, and chopper pilots to whisk us and our loved ones to safety? What happens when tapped-out insurance companies can't pay home owners for their losses? Or when Americans are worn out by disaster fatigue and there's no national outpouring of donations and resources to nurse a mortally wounded city back to life? "Katrina-like disasters could become commonplace along vast stretches of U.S. coastline in the not-so-distant future," noted Mike Tidwell in *The Ravaging Tide*, which deals with the coming crisis in coastal cities as the planet warms. "And evacuating inland might not be an option, no matter how bad the storm, because extreme weather events in the heartland (drought, heat waves, forest fires) will remove the welcome mat. There simply won't be the infrastructure and surplus resources needed to absorb the overflowing humanity."

"We all lull ourselves into a sense of complacency assuming everything is being taken care of and think, *If this went to hell in a handbasket, the federal government would rescue me in 24 hours,"* DeBlieux, the ER chief who was stranded at Charity Hospital, told me one morning while we sat in the living room of his rebuilt woodsy craftsman-style home at the edge of New Orleans's City Park. "I thought that was reasonable, but I'm living proof that doesn't happen. I don't have faith in the existence of a safety net from the federal government anymore. Five days into this and we still weren't evacuated, and I'm the health care provider taking care of people in a hospital. That's absurd in anybody's book. Seventy-two hours is absurd. And this will occur somewhere else, whether it's in Florida, or Alabama, or Texas. It will happen again."

DISASTER-PROOFING HEALTH CARE

"You need to do more stringent monitoring and calibrate your meds," said Felice Guimont, the former ER nurse from Charity Hospital who is now a nurse at the New Orleans Musicians' Clinic (NOMC), during a monthly luncheon sponsored by the organization. "But exercise and diet are just as crucial."

She was talking about the importance of healthy lifestyle habits in a matchless setting: the spacious enclosed patio of a stately, whitewashed antebellum mansion with colonnades and sweeping verandas in New Orleans's historic Garden District. About two dozen musicians and friends listened intently as Guimont dispensed tips and diabetes-friendly recipes while they munched on Cobb salads, vegetarian wraps, and chocolate chip cookies inside the greenhouse. Floor-to-ceiling louvered windows offered ventilation and overgrown ficus, palm, and bamboo trees brushed the rafters on the house's second floor. In an adjoining antique-filled room furnished with Tiffany lamps and Victorian wicker chairs, staffers working at computer stations ran the command center for the clinic.

It was a steamy June day in 2012, and Guimont kept cool in a gauzy blouse and jeans. A musician herself, she understands how difficult it is to keep blood sugar in check when leading an itinerant life. Many musicians live hand-to-mouth existences, work odd hours, are frequently on the road, and neglect their medical care for lack of health insurance. The clinic, which is funded by grants and private donations, provides free or reduced-cost health care to musicians and was cofounded 15 years ago by the house's owner, Bethany Bultman. She was astonished and alarmed to discover that so many musicians—who are such an essential part of New Orleans's unique culture—were dying young of easily preventable diseases mainly because they never went to the doctor.

The clinic itself is located nearby, in the heart of the Garden District, in a suite of offices and a couple of exam rooms on the second floor of a medical building owned by Louisiana State University. There, more than 2,800 musicians are treated for everything from the flu to hypertension, depression, heart disease, diabetes, and even hepatitis C. It offers a full range of services, including education, treatment, and prevention, that are geared toward musicians, whose lifestyles can cause serious health problems such as hearing loss, carpal tunnel syndrome, lung diseases from smoking and breathing second-hand smoke, and often depression from working in such a financially tenuous profession. A social worker helps them navigate the health care system if they need to see specialists, dentists, or pharmacists, who charge NOMC clients reduced rates. When Guimont herself was stricken with diabetic retinopathy, which is an irreversible cause of blindness, she lacked health insurance. "But I went to the clinic," she recalled. "They saved my eyes."

The clinic is part of a renaissance in medical care that, ironically, has positioned New Orleans at the cutting edge of reforming our grossly inefficient and wasteful health care system. What they're doing here and in other parts of the city is an entirely new way of tending the sick that can not only prevent illness, but also be more resilient and flexible in a disaster. In the ruins of the decimated public health system immediately after Katrina, a group of local doctors set up 18 makeshift street clinics in the city's poorer neighborhoods to treat the indigent walking wounded. In the years since, their grassroots effort has proved so successful that it has evolved into a network of more than 100 community-based clinics run by 25 different organizations. These are supported by a combination of government funds, private grants, and charitable donations from organizations like the New Orleans Musicians Assistance Foundation, and they handle 175,000 visits a year from about a fifth of the city's residents, the vast majority of whom are uninsured. This network is now widely hailed as a new medical paradigm for delivering health care where it is most needed.

"Katrina gave us a clean slate—an unprecedented opportunity to rebuild a broken system from the ground up," said Dr. Karen DeSalvo, who helped establish these clinics and is now the health commissioner for the city of New Orleans. "We always thought that if a storm flooded the city, it would be 'game over' and we'd all pack up and leave. But something very different happened and made us all want to fight for the city. The work that has happened here to rebuild and restructure our community is one of the finest examples of civic engagement and people defining their future and developing their own laboratory of innovation and change."

Even before Katrina, New Orleans was riddled with profound poverty, and residents suffered from exceedingly poor health that was worse than that in many developing countries. Louisiana ranked first or near the top on a long laundry list of benchmarks, including the prevalence of smoking and obesity; deaths from heart disease and cancer; rates of infant mortality, asthma, and diabetes; preventable hospitalizations; and percentage of people without health care insurance. Roughly one out of every four people in New Orleans—about 22 percent of the population, or 100,000 people—was uninsured before the storm, compared to about 18 percent nationally. African American males in New Orleans had the highest incidence of lung cancer in the world. Many of these health conditions were simply the result of bad habits, poor diets, and lack of exercise. When DeSalvo was on staff at Charity, she saw a lot of patients die of illnesses like breast cancer or end-stage syphilis and tuberculosis that could have been easily cured or at least controlled if they had seen a doctor sooner. Later, as head of the department of internal medicine and an administrator at Tulane University Hospital, one of the questions that constantly percolated in her head was how best to reach these patients, because it was clear that current methods of caring for the indigent in public hospitals and emergency rooms was grossly inefficient, costly, and simply ineffective.

"We took matters into our own hands, because there wasn't a coach or a playbook," she recalled in an interview in a no-frills conference room at City Hall, where the health department has its offices. "This evolved from the largely street-based efforts that landed in a church, a dormitory, a grocery, or Lord and Taylor, and from our experiences right after Katrina, at urgent-care stations with ice chests with vaccines in them and volunteers scurrying around working at card tables, and the understanding that to make our community healthy we'd have to meet people where they were. If we gave a patient insulin, that doesn't help if they don't have a house or a place to refrigerate it. We had to build a neighborhood-based primary care system and move away from a system that was all concentrated downtown. Prior to Katrina, it was highly centralized financially and geographically. It was a model of care that was not at all flexible in the face of disaster. We had all of our eggs in one basket, which is a mistake."

Instead of relying on large, public hospitals to be the front lines for the uninsured, these neighborhood-based "medical homes" offer primary care, screening tests, mental health services, and referrals to specialists. Doctors and nurses can build relationships with patients—and manage or even prevent debilitating and deadly ills like asthma, diabetes, and heart disease that can be well controlled with regular attention—and provide the kind of comprehensive care that is simply not available in emergency rooms.

Moving these basic safety-net services from downtown New Orleans, where they were heavily concentrated in the medical complex that encompassed University and Charity Hospitals, into local areas means patients can walk to clinics instead of taking several buses. This makes it easier for them to get and keep appointments and to have the continuity of care and follow-up that are not possible in a hospital. The total cost savings with using this approach are enormous, in the millions of dollars—it costs about $600 a year to treat a patient at a local clinic versus about $600 for one emergency room visit—plus, the care patients receive is better and they have more confidence that they will receive high-quality care, according to a recent Commonwealth Fund study. In a 2009 survey of 1,573 patients at 27 of these clinics, they found that only 27 percent of these respondents went without needed medical care because of cost compared to 41 percent of adults nationwide.

In addition, patient records are now kept electronically, and New Orleans is also part of a federally funded pilot program that stores this information in a central information exchange. "That way, in a disaster, the data follows the patients," said DeSalvo. "And instead of having a single safety-net institution, we have 25 organizations that work together to provide primary care and mental health services in the four-parish area. What that means is that if one of those providers gets overwhelmed, they can call their friends—literally—and they have."

Outside the window of DeSalvo's eighth-floor office, a revolution of another sort is happening. Construction is proceeding on a 70-acre medical campus downtown, a combination academic and Veterans Affairs medical complex with a $2.4 billion price tag that will replace Charity and a VA hospital that was also damaged by Katrina. The two new hospitals, which will have more than 600 beds in total, are slated to be completed by 2015. While the decision to shutter Charity Hospital permanently came under fire from neighborhood activists who wanted to preserve the landmark institution, supporters believe the new complex will revitalize the city's core and put New Orleans on an equal footing with the highly successful medical corridors in Houston and Birmingham, Alabama.

"Charity lost its entire central nervous system because all of its information technology was in the flooded basement," said Dr. James Diaz, director of the environmental and occupational health sciences program at the LSU School of Medicine. "So why would you want to retrofit a building that's 70 years old? That doesn't make sense. This is the biggest medical complex under new construction probably just about anywhere in the South and it'll be the largest employer in the city. This is an opportunity to build a first-class medical center that has everything we need right on campus and that truly is on par with Birmingham and Atlanta and Houston."

But DeSalvo is the first to admit it will be a "generation" before New Orleans's health care system completely recovers; there are places in the city that bear deep psychic and physical wounds that may never heal.

"My family lived here for generations, and they were dispossessed," Felice Guimont told me as we drove along the bumpy streets she rode her bicycle on when she was growing up and attending a nearby parochial school in the Lower Ninth Ward. The mostly African American neighborhood along the Mississippi sustained the hardest hit when the levees broke because of two breaches in the adjoining Industrial Canal. Once, this was an orderly, thriving community of more than 18,000 people with comfortable single-family homes and well-tended lawns, but less than a third now remain in the two-mile-square district that has become a jungle of weeds, trees, and 12-foot-tall grasses and a home to roaming packs of stray dogs.

Sure, there are a dozen or so of the new eco-friendly, pastel homes with the angular, modernistic lines of Scandinavian architecture built by Brad Pitt's Make It Right organization. Rebuilt brick bungalows with neat fences are scattered along the rutted roads. But vast stretches of land are still littered with the wreckage of abandoned homes and shattered lives. Some lots are gigantic trash heaps for worn-out tires, construction debris, and the sodden refuse of the city.

"Most of these houses were completely underwater," said Guimont, pointing to the gutted lots that are now covered with vegetation. "That was where my best friend's house was—it used to be so beautiful—and Fats Domino's house was over there," she continued. Soon we stopped in front of a large rectangular chunk of land bordered by wildly overgrown hedges and blanketed with tall grasses. "This is where my parents' house was," she said quietly, her eyes clouding over. "Every Sunday, I think about having dinner with my family after church and then I remember that none of them are here. I can't keep track of all the people who died afterwards from the stress of the dissolution of families and the fragmentation. I went to a high school football game at my son's alma mater not long ago—he graduated in 2004. Everyone looked so old and some people's hair had turned white. It was like a science fiction movie. This city is still packed with grief."

CHAPTER FIVE

RUNNING ON EMPTY

F amine has plagued humanity throughout history. But as increasingly frequent and severe droughts parch farmlands and decimate food supplies, climate change may make hunger and malnutrition harrowing facts of life for growing numbers of people throughout the world. The total area of Earth's surface classified as "very dry" has doubled since the 1970s, and within the next two decades, nearly half the world's residents will be living in regions of severe water stress, according to recent studies, which will inevitably impact the food supply chain. These calamitous events have caused widespread suffering and death since ancient times. But what we haven't known until relatively recently is the terrible toll famine takes on the survivors in the long term.

In the past few years, scientists have deepened their understanding of the lingering physical and psychological consequences of malnutrition by studying an episode in Dutch history known infamously as the Hunger Winter. This tragic period occurred in Nazi-occupied Holland at the end of World War II, over the winter of 1944–1945. Millions were pushed to the edge of starvation and thousands died. Because the Dutch kept such meticulous records, the Hunger Winter provided epidemiologists with the details of a ghastly living laboratory that revealed the dire effects of starvation. Virtually no one living in the Netherlands at the time was insulated from the fallout of this catastrophe, including the most famous victim of this period's dreadful privations, the actress Audrey Hepburn, whose private battles with ill health throughout her later life reflected the physical traumas that plagued her compatriots who also endured this ordeal.

The Belgian-born actress was one of the biggest screen stars of the '50s and '60s and endowed a series of iconic roles, from Holly Golightly in *Breakfast at Tiffany's* to *My Fair Lady*'s Eliza Doolittle, with a mixture of regal elegance and spirited mischief that made her an enduring audience favorite. The dark-haired beauty had the erect carriage and fawnlike grace of a classically trained ballerina, a radiant 1,000-kilowatt smile, and a captivating but indefinable accent that could be traced to her European roots as the daughter of a Dutch baroness and a wealthy English banker. Her father walked out when she was just six, but her early life was one of privilege, studying ballet and attending posh boarding schools in London. But when she was trapped as an adolescent in her native Holland after Hitler's army invaded in 1940, Hepburn and her family suffered harsh privation: Her brother was dragged off to a labor camp, another brother went into hiding, and her uncle and a cousin were executed for aiding the Dutch underground. She lived out the war in her grandparents' home, which, coincidentally, was a few blocks away from a warehouse where another teenage girl her age was hiding in the attic: Anne Frank.

But the Oscar winner's delicate, waiflike loveliness was born of the starvation she experienced in the waning days of the war, which may have caused genetic changes that permanently impaired her health, according to new research by Dr. Nessa Carey of Imperial College London, author of *The Epigenetics Revolution*. After the Allied landing on D-Day in 1944, Germans blockaded food and fuel supply routes in retaliation for national railway strikes called for by the Dutch government in exile in London. In the darkest days of the embargo, the entire population in western Holland, which included Amsterdam and comprised more than 4.5 million people, was trying to survive on about 30 percent of its normal daily caloric intake. By April 1945, official rations had dropped to 500 calories per day. City dwellers would make long pilgrimages to farms in the countryside, walking for miles to trade anything they could carry for a crust of bread. They burned every piece of firewood and furniture they could get their hands on just to stay alive. Yet even when the blockade was lifted, it was still difficult to get goods through because the canals had frozen. Nearly 30,000 Dutch perished from starvation and the bitter cold by the time supplies were restored in May 1945, when the Allies liberated the region and US bombers began food drops in the Netherlands.

Like her neighbors, Hepburn, then 16, was often forced to subsist on grass and tulip bulbs, and she holed up in a cellar in the war's final days. When she emerged from hiding, she was sickly, stricken with jaundice, respiratory problems, and what would become chronic anemia. Her spindly ankles had become so swollen that they derailed her once-promising dance career. For the rest of her life, Hepburn quietly coped with fragile health—migraines, several miscarriages, and, in her later years, clinical depression—until her untimely death at age 63 from a rare intestinal cancer. Many of those who survived, like Hepburn, endured lifelong health woes that likely stemmed from months of malnutrition.

Scientists attribute this to epigenetic changes. Epigenetics is a relatively new field that studies how what's encoded in our DNA can be changed by our behavior, our surroundings, or traumatic events. "Life histories become altered irrevocably in response to their environment, causing imprinted genes to switch off at certain stages in development and stay off throughout the rest of life," explained Dr. Carey in a recent phone interview from her home in the English countryside. In Hepburn's case, "because of the malnutrition she suffered as a teenager, it caused subtle modifications in the way her genes expressed themselves."

The special circumstances of the Dutch Hunger Winter gave scientists unprecedented access to data they needed to study the effects of starvation. Because the Netherlands was an advanced and well-structured society and people were well fed before and afterward, the famine had a clear beginning and end and everyone was affected equally, not just a troubled few. During the famine, Dutch authorities also maintained thorough health care registries and food-rationing documentation, which elucidated in gruesome detail how conditions deteriorated over time. This provided researchers with a rare opportunity to study the damage done by malnutrition in ways they hadn't been able to in other populations that endured similar traumas, such as the food deprivation that occurred during the Dust Bowl years here in the United States, the seasonal famines in Africa and Bangladesh, the Siege of Leningrad in the early 1940s, and the Chinese Great Leap Forward famines in the late 1950s and early '60s, which claimed as many as 30 million lives and led to cannibalism. The Dutch experience provides an unfortunate glimpse into the long-term health consequences of starvation, and perhaps into the world of the future.

In the early 2000s, a joint Dutch–American research team tracked down many of the surviving 60-year-olds who had been gestating in their mothers' wombs during this terrible time. Because the Dutch kept such excellent records, they were able to locate and obtain blood samples from close to 2,400 adults who were fetuses just before, during, or after the Hunger Winter. Each individual's sample was compared with that of a same-sex sibling born at a different time. Whether subtle genetic changes could be correlated with long-term health consequences was "the million-dollar question," said epidemiologist Lambert H. Lumey of Columbia University, who collaborated on the study. But it soon became abundantly clear that this monstrous ordeal left a permanent scar on their DNA. "Early-life environmental conditions can cause epigenetic changes in humans that persist throughout life," the researchers noted. Across the board, children born during this period had higher rates of heart disease as they aged than their peers did. The fetuses whose mothers went hungry in the first trimester of their development were harmed the most: As adults, they were more likely to be obese or overweight, as if some hoarding gene had been permanently turned on in utero, and they suffered more often from diabetes, high cholesterol, and clotting disorders. Malnutrition severely impacted their neurological development, too. These youngsters experienced twice the rates of schizophrenia and were stricken more frequently with mood disorders such as depression later in life.

Today, more than two billion people around the globe, in the developing world and even here in the United States, live on the ragged edges of survival and experience hunger, which makes them not only more susceptible to infectious diseases because of their weaker immune systems, but also more vulnerable to these patterns of long-lasting genetic damage from malnutrition. Their ranks will only swell as the world becomes more environmentally stressed. Within the next 20 years, food prices could more than double, according to Oxfam, the international relief group, with climate change being the culprit for at least half of that increase. By century's end, rising temperatures could spark drought-induced food shortages for half the world's population, Stanford University researchers have projected.

And that study looked just at the effects of withering heat—researchers didn't factor in the effects of water scarcity, which would worsen an already tragic situation.

In the absence of another green revolution, we could be looking at mass starvation on a global scale. "By 2030, we could experience real food shortages in Southeast Asia and southern Africa," David Lobell, a member of Stanford's food security research team, told me in a recent interview at his office on the Palo Alto campus. "If you look at scenarios 60 to 70 years out, when temperatures could rise three or four degrees, things look very bleak: There will be less water, and less rainfall in major crop-producing regions, with groundwater being depleted faster than it is replenished by rainfall, and [water for] irrigation not being available and much higher swings in temperature from year to year as the climate becomes more erratic. The outcome could be that hundreds of millions more people are pushed into hunger" in the developing world and even in newly emergent industrialized nations like China, Indonesia, and India.

Climate-induced food shortages will almost inevitably lead to political instability, mass migrations of millions, and even civil war, and Americans won't be shielded from the fallout. While much of the current volatility in Africa and the Middle East is attributed to religious extremism, ethnic divisions, and political factions, the fact is that impoverished people who don't get enough to eat tend to become incensed about their plight. Systemic poverty, increasing food insecurity, unemployment, and water scarcity are all what defense experts call "threat multipliers"—underlying factors that amplify existing divisions and set the stage for violence. Soaring prices for staples after floods and droughts caused widespread crop failure in 2007 and 2010 were partly to blame for food riots in India and Mexico, and may even have ignited the spark that led to the Arab Spring. A team led by Marshall Burke, an agricultural economist at the University of California in Berkeley, found there was indeed a connection between rising temperatures and violence. They looked at African wars from 1980 to 2002 and found that civil wars were more likely to begin during hotter years, and that a 1°C rise in temperatures was linked to a 50 percent increase in conflicts in that year.

"We did a pretty simple analysis," Burke told me while we sat in his sparsely furnished office in one of the magisterial granite art deco–style buildings on the Berkeley campus. "We took existing data sets of historical rainfall and temperatures and looked to see if there was a connection to civil conflicts. Then we took the records by country, by year, and by whether or not they had a civil war. The link between temperature and violence was striking, even more so than precipitation. What seems to be driving this relationship is the drop in crop yields, which are really sensitive to small changes in temperature. When yields go down, economic opportunities more broadly go down, which provides the impetus for them to join rebellions."

Because temperatures are expected to go up by at least 1°C (about 1.8°F) over the next 20 years, Marshall and his colleagues estimated that climate change could increase the number of civil wars in Sub-Saharan Africa over that time period by 55 percent. The human costs could be staggering, with at least 400,000 lives potentially being lost in these conflicts. "And these are just the battle deaths, which account for only about 10 percent of mortality during conflicts," Burke told me. "The likely toll will be an order of magnitude higher—possibly four million—because of people who are displaced and malnourished, outbreaks of cholera, or other disease epidemics."

FARMING IN HELL

Although recent droughts have been disastrous for ranchers and farmers in the United States, Americans still remain somewhat insulated from the full force of climate change and the projected food shortages. But not for long: Agricultural areas throughout the Midwest and especially in California are threatened by the growing water insecurity that accompanies a warming world. We're not doing nearly enough to avert disaster. The torrid summer of 2012—which came at the tail end of the warmest 12 months in the contiguous United States since record keeping began in 1895—may be just a preview of the kind of dry conditions we'll see in the farming heartland as the world warms. The consistently triple-digit temperatures sucked up moisture from the air and the soil, creating a severe drought in more than half of the United States. It was like a slow-motion train wreck—a "creeping disaster" in meteorological terms—for farmers in America's Corn Belt. Corn reproduction is fiendishly tricky, requiring intricate maneuvers worthy of a Cirque du Soleil contortionist: A grain of pollen must alight on one of the wispy silken threads that surround the husk and then somehow work its way inside the husk to pollinate a single kernel. And corn is notoriously heat sensitive—once the thermometer rises above about 84°F, yields drop by up to 30 percent. "You couldn't choreograph worse weather conditions for pollination," one crop biologist told *Bloomberg News* in July about 2012's double-barreled punch of sizzling temperatures and parched weather. "It's like farming in hell."

Little wonder, then, that only 26 percent of the corn crop that summer was rated good or excellent, according to the USDA, which designated as disaster areas more than 1,000 counties in 26 states from South Carolina to California, including much of the South and Midwest, the largest such declaration in its history. The crop failures prompted corn prices to rise more than 40 percent, and the poor yields and the bad weather had a ripple effect throughout the entire food chain. Livestock farmers were forced to buy high-priced feed because their pastures were seared, some desperate ranchers were forced to sell off cattle early because they couldn't afford to feed their herds, and bone-dry land that had been in families for generations was threatened with foreclosure.

The record-low rainfall reduced water levels in the Mississippi River to such a dramatic degree that sandbars emerged midstream, and the edges of the green riverbanks were lined with broad sandy beaches showing where the water had receded. In some places, traffic virtually stopped on the river, a transportation corridor that normally carries 60 percent of the country's grain. At one point during the long, hot summer, grain elevators overflowed with a huge backlog of millions of bushels of corn sitting under tarps, waiting for barge space. Much of the soybean crop spoiled while sitting in storage silos upriver.

The shortfall translated into a cascade of higher food costs in the United States, not just for grains, but for dairy products and beef, too. In America, only about 13 percent of household budgets are spent on food, according to the Bureau of Labor Statistics. But lower production here in the United States translates to higher prices and less supply abroad. Consequently, the real misery was felt most deeply by the world's poor in the developing countries, where a billion people suffering from hunger were forced into more profound privation because, with as much as 75 percent of their incomes already going for food, according to Oxfam America, any price hike means they must do without.

But with climate change, we're facing a much drier future, and one for which we've made scant preparation. "Look at what we're seeing right now," said Glen MacDonald, a geographer and director of the UCLA Institute of the Environment and Sustainability. "The preponderance of the evidence is that global warming is building at a rate faster than many of us thought it would. I think the nation has tremendous capability both in terms of technology and intellectual capability to marshal our resources and to deal with challenges. But we can't hobble our economy and our agricultural productivity by putting our head in the sand—I don't think that's a responsible course of action."

Clean water is the lifeblood of modern society. None of the incredible scientific advances of the 20th century have affected human health as profoundly as the availability of clean water. Cholera, typhoid, and malaria have virtually vanished in industrialized nations because of municipal sewage and drinking-water systems. And clean water has helped to feed the world: Irrigation for agriculture accounts for at least two-thirds of all water use. Global warming may change all that. Societies historically have adapted to changes in weather and climate by diversifying crops, altering irrigation methods, and even devising crop insurance to survive in the bad years. But a hotter planet could create conditions that are beyond the scope of our typical coping mechanisms. By the end of this century, nearly half the world's population—about 3.2 billion people—could be enduring severe water scarcity.

Human history is inextricably entwined with water because patterns of settlement were dictated largely by the availability of this precious resource. In the United States, the 20th century was marked by bitter battles over water rights, endless stories of small farmers being swindled out of their birthrights by greedy land barons and corrupt civic leaders, the migrations of more than two million Americans from the drought-stricken Dust Bowl, and the heroic engineering projects built in the depths of the Great Depression that dammed mighty rivers and funneled millions of gallons of water through massive aqueducts to remote desert outposts, leading to the phenomenal expansion of the American Southwest and facilitating a huge migration from the North to the Sunbelt.

Paradoxically, we're not actually running out of water—we have the same amount of water on Earth now as we did eons ago, when amphibians first stuck their toes out of the drink and explored dry land—about 360 quintillion gallons of it. The hydrologic cycle, which we learn about in grade school, still follows the laws of chemistry: Water evaporates from the oceans, lakes, and rivers, then forms clouds in the atmosphere, falls as rain, and seeps into the ground before surfacing once again in springs that flow into rivers and lakes. But the problem now is that we're using it and polluting it faster than it can be renewed—hence the dire shortages, a situation that Peter Gleick, president of the Pacific Institute and one of the world's leading hydrology experts, has dubbed "peak water," which he describes as a crucial tipping point when nature's ability to replenish stores of freshwater is forever outstripped by voracious demand fueled by explosive population growth.

Worse, "there are now indications that the hydrologic cycle is accelerating, which is exactly what climate modelers and hydrologists said would happen from a warming planet," Gleick told me when we talked in his office at the institute, located in a blue-and-gray restored Victorian home in Oakland, California's preservation district. "That means more evaporation, more precipitation, more runoff into the ocean, which translates into dry regions, like the Southwest, getting even drier because of higher evaporation rates and wet areas getting inundated by floods because of faster melting of snowpack. That's part of the new climate we're going to have to deal with. I think we're already unavoidably committed to some negative impacts of climate change because we've waited too long. We can move toward sustainable water management—and we will—but the question ultimately is: How much pain are we going to suffer because we're moving too slowly?"

What we're seeing is what hydrology experts call the end of "stationarity—the idea that natural systems fluctuate within an unchanged envelope of variability," according to Gerald Galloway, a civil engineering professor at the University of Maryland and a leading expert on water resource management in a warming world. Climate change has thrown a wrench in our usual projections, which makes predicting our water futures exceedingly tricky, Galloway told me during a recent interview. He ticked off a long list of water challenges that face America: more frequent and severe droughts; rising water demand from a population that is expected to grow by up to

150 million people by 2050; increasing flood damage; inadequate protection of groundwater, vital wetlands, and the environment in general; aging and inadequate water infrastructure; lack of adequate watershed planning; and on and on.

California is, unfortunately, exhibit A in the bureaucratic stalemate that has crippled meaningful water reform. The raging battles that have erupted in the Sacramento–San Joaquin River Delta, the headwaters of the two rivers that supply the state with much of its freshwater, are emblematic of the water-rights fights taking place across the country. But torrential rains, droughts, or an earthquake causing the collapse of this waterway system— which is already near the breaking point—could spell disaster for the nation's food supply because the Golden State's $40 billion agriculture industry produces nearly half of our domestically grown fruits, nuts, and vegetables.

Russell van Loben Sels has been at the forefront of these skirmishes for more than two decades. One temperate October afternoon, I visited him at his family's farm on the Sacramento River, which forms a natural border for some of the acreage his family has owned for 130 years. Tall and sinewy with the leathery face that comes from a lifetime outdoors, the 60-something farmer has spent most of his life on the delta, which is a verdant estuary just south of Sacramento. At Amistad Ranches, which is what their farm is called, they grow wine grapes, safflower, alfalfa, tomatoes, and pears on 2,500 acres in Walnut Grove and Freeport, two tiny delta towns. "If you look at the delta," said van Loben Sels, gesturing toward the river, "almost every drop of water between us and the foothills of the Sierras comes right in front of our front door here."

Van Loben Sels left the area in the early 1960s to attend Stanford University in the leafy glades of Palo Alto about 80 miles to the west, where he majored in economics. After an Air Force stint and a brief foray into the corporate world, he returned home, a place light-years removed from the trendy enclaves of the San Francisco Bay area where he attended college. About half a million people live in the delta, in cities like Sacramento and Stockton and in more than a dozen rural hamlets along the meandering river. The area has the squat scrub and languid vibe of Louisiana's bayou country. These parts of California still seem frozen in a post–World War II time warp with their weather-beaten motels, old-fashioned diners, and all-purpose general stores.

The delta itself is a triangular wedge of land sliced up into a checkerboard of green and brown fields, through which thread the spidery fingers of serpentine rivers and swamps. This was once a 700,000-acre marshland that Gold Rush settlers dredged and tamed with what eventually evolved into a 1,300-mile network of levees to create the artificially dry islands that have been farmed for more than a century. Van Loben Sels's great-grandfather, who emigrated from Holland and carried with him a deep understanding of his native land's success in carving land from swamps, helped build many of the delta's original earthen dikes to protect crops from the river. But much of the land has settled in a process known as subsidence, which was accelerated by the continuous cultivation of the rich black peat, and most of these islands are now more than 20 feet below sea level.

The levee system was jerry-built, made with mostly mud and mortar barriers and rudimentary pumps to dredge the marsh, and it worked well enough when it chiefly served local farmers. But the delta is also the heart of California's freshwater system, a vast hydraulic switchyard where the Sacramento and San Joaquin Rivers converge before emptying into the San Francisco Bay, and it supplies two-thirds of the state's water. At the delta's southern edge, two gigantic pumping stations with more horsepower than 100 diesel locomotives literally reverse the rivers' flow and divert the water southward. Then it is channeled through a convoluted network of hundreds of miles of canals, tunnels, and aqueducts to the taps of 28 million Californians and the irrigation spigots for five million acres of farmland in the San Joaquin Valley, which supplies half the nation's fruits and vegetables and one-quarter of its dairy products.

Starting in the 1930s, when engineers began harnessing the delta's waters to quench California's growing thirst—and transforming the dusty Central Valley into the world's richest agricultural empire and fueling unbridled urban growth—the delicate ecology of the river system began a long slide into collapse. Fueled by the ravenous demand, those massive pumps suck up fish vital to the habitat's food chain and upend the natural tidal ebbs and flows that replenish the web of waterways. "The current operation of the delta is simply not sustainable—even if a miracle occurred—because we're taking more out than this ecosystem can sustain," said van Loben Sels, who heads the Farm Bureau Delta Caucus, a coalition of five county farm bureaus that aims to provide a unified voice for agricultural interests in California. "There really isn't an unlimited supply of water."

Climate change could be the death knell. While farmers here tend to be politically conservative—a spot of red surrounded by a sea of blue in one of the nation's most liberal states—you'd be hard-pressed to find global warming skeptics, because most of them have seen the effects of a change in the weather firsthand. And they're afraid—of floods from heavier rainfall, of the disappearing snowmelt from the rugged western slopes of the Cascade Mountains and the Sierra Nevadas that provides the estuary's storehouse of freshwater, and of rising seas, which are expected to swell by at least three feet in the next few decades. Violent megastorms, the kind that normally occur only once every 300 years, could become more frequent and powerful because of global warming. Squall systems like the Pineapple Express, which sucks up water from the ocean around Hawaii and pummels the mainland with drenching rains and fierce winds and waves, could trigger massive floods and mudslides that wash away up to 25 percent of California's homes and overwhelm the state's flood-protection system. The last megastorm happened in 1861, when California was battered with 45 days of such severe storms that the Sacramento Valley became "an inland sea," according to the US Geological Survey. The deluge caused so much damage that the state capitol was moved temporarily from Sacramento to San Francisco and Leland Stanford was forced to take a rowboat to attend his gubernatorial inauguration.

Any one of these threats—along with the ever-present risk of a major earthquake along at least half a dozen faults in the San Andreas zone near the delta—could lead to the catastrophic breakdown of the makeshift levee system and trigger what hydrology experts have called the Big Gulp: ocean water being sucked in from the Pacific, doubling the size of San Francisco Bay, inundating the delta, filling up below-sea-level fields like bathtubs, and submerging vast swaths of fertile farmland as far east as Stockton in salty water. "It doesn't take a rocket scientist to figure out what is going to happen," van Loben Sels sighed. "The river set the water table for this whole region. If it's five feet higher because of the ocean, we're going to have fields we can't farm because it will be too wet."

Within the next 40 years, experts predict there is a 64 percent chance of a large-scale rupture in the delta region due to either an earthquake or a megastorm. These are sucker's odds—roughly a two out of three chance that a massive failure will trigger a disastrous cascade that washes away critical infrastructure for much of the state: the highways, railroad tracks, gas lines, aqueducts, telecommunications networks, hydroelectric power grids. Millions of acres of land and cities from the Silicon Valley to San Diego would be cut off from their major source of water for months, and possibly even years. Sacramento, where the Sacramento and American Rivers meet and the United States' most flood-prone city, would be swamped under up to 20 feet of water. Such an event would create far greater havoc than Hurricane Katrina in terms of the number of deaths, amount of destruction, and overall cost, according to Robert Bea, a professor emeritus of engineering at the University of California, Berkeley, and a leading expert in catastrophic risk management; it would cripple the state economically and cause a depressing ripple effect throughout the nation and the entire global economy.

The blunt-spoken civil engineer has spent more than a half century investigating high-profile engineering failures, from the New Orleans levee breaches after Katrina to the Columbia space shuttle disaster and the explosion of the *Deepwater Horizon* offshore drilling rig in the Gulf in 2010. He even lost his own home when Hurricane Betsy flattened New Orleans in 1965. Now in his 70s, his voice is raspy from a recent infection, but his critical faculties are undimmed. He heads up a $2 million project funded by the National Science Foundation called Resilient and Sustainable Infrastructure Networks (RESIN) that is charged with coming up with a blueprint to safeguard the imperiled delta. What he's uncovered so far is not comforting. "We wanted to evaluate what engineers call the 'true risks' associated with our delta infrastructure, recognizing the interactivity and the interconnectivity," he told me during an interview at his home in Moraga, a woodsy suburb east of Berkeley. "And what we found was that the risks were so far out of the acceptable regions it makes you sick. We haven't had an earthquake in the delta area for a hundred years. The energy is building up. The longer you let it build up, the worse it is going to be when it pops."

And unfortunately, like Humpty Dumpty, once the delta fails it will be next to impossible to put it back together again, experts have said, because the levee system is so intricate. This impending threat has reignited the water wars that have simmered in California ever since William Mulholland, the first head of Los Angeles's Department of Water and Power, poached water from the Owens Valley in the Eastern Sierra Nevada a century ago, which inspired the movie *Chinatown* and set the stage for the transformation of Los Angeles from a chaparral desert into a world-class city. The way this precious resource has been allocated in California—and in most of the West—has been on a first-come, first-served basis: Whoever signed up for water rights contracts first got first dibs on as much water as they wanted—namely, the northern delta farmers like Russell van Loben Sels and his family, who've been here since the late 19th century. Relative newcomers in the San Joaquin Valley got junior rights, which meant they would lose at least some of their shares during dry spells. But over the past three decades, water shortages have become chronic and droughts more frequent. Lawsuits filed by environmental groups to protect endangered habitats, along with government-imposed regulations that require compliance with new federal ecosystem standards, have meant that increasing amounts of water have been set aside for wildlife restoration. As a consequence, farmers in parts of the valley have been forced to buy water from their eastern neighbors or rely on pumping it from underground aquifers, which are rapidly being depleted.

With a changing climate, the handwriting is on the wall. In 2009, then-governor Arnold Schwarzenegger and the California legislature passed a historic package of bills for managing the waterways and established a governing body, the Delta Stewardship Council, that brought together a diverse group of players, including environmentalists, municipal water agencies, delta farmers, big agricultural interests, and developers who require water for the housing tracts that will be needed to shelter the estimated 20 million people California is expected to add in the next 20 years. The centerpiece of the delta revitalization plan is a proposal to replace the antiquated plumbing system with a canal or tunnel that will divert freshwater directly from rivers and funnel it to communities throughout California, bypassing the delta altogether. California voters resoundingly rejected a similar plan for a $1.3 billion peripheral canal more than 30 years ago, in 1982, over the fears of residents in the northern part of the state that those to the south would suck up all their water. But the water situation has grown grimmer in the years since and the state's population has jumped by 10 million. Water managers, environmentalists, and other experts now believe the canal would provide a more dependable supply of water, and restore fish and wildlife. "What we're doing in the delta now isn't working," Peter Gleick told me. "The delta's broken. And a big part of what's broken is the way we take water out of the delta, the place we take it, and the amount of water we take out. If we continue business as usual, the ecosystem will collapse, salmon will go extinct, the water quality for Southern California will worsen, and the reliability of deliveries to all of the users south of the delta is going to get worse. Everyone's going to suffer."

In 2012, governor Jerry Brown resolved to push through the peripheral canal, which would consist of two 35-mile-long tunnels to tap into the Sacramento River north of the delta—asserting with audible frustration, "I want to get shit done." But the project still faced strong political head-winds and assorted groups, including environmentalists, consumer advo-cates, tribal leaders, fishermen, and family farmers, remain convinced the canal is just a $50 billion *Chinatown*-style boondoggle engineered by wealthy Southern California developers. Russell van Loben Sels wants to protect the legacy his family and his neighbors have worked for genera-tions to create, and they fear their irrigation water will grow saltier as freshwater is diverted, that portions of their land will be gobbled up by the new infrastructure, and that they'll be robbed of even more water. "What's being proposed is terrible," he said, going on to complain that several closed-door meetings of the council had pointedly excluded small farm-ers. "I'm concerned what is happening will have lasting effects on the delta, not just on my water supply. That ecosystem needs to be protected."

The plan continues to languish in limbo, a seeming victim of governmen-tal gridlock. Indeed, when Robert Bea met with California's federal law-makers in Washington, DC, to share his initial observations about the severity of the crisis, he says he was met with blank stares. "Their response was, If the risks are unacceptable, what can be done about them?" he recalled. "The problem is that this requires long-term solutions—it can take up to 40 years to erect new waterway structures. We've developed this infra-structure over 200 years and it can't be resolved within the next election cycle. Can we work this problem out in a responsible way or do we wait until it fails—will we fix it now versus pick up the pieces later? If we wait, it will be 10 to 100 times more costly to make the repairs. But our politicians can't come to grips with it, so nothing is being done and we're at a stalemate."

DEAD POOLS

It's not just farmers who are feeling the pinch. Climate change could dry up once-thriving Sunbelt cities that flourished because engineering feats shunted water to these parched territories. But out of sheer necessity, the driest city in our driest state has learned to cope with scarcity: Las Vegas, a desert oasis brought to life by a visionary gangster in the heady years after World War II.

His underworld pals called him "Bugsy" because they thought he was crazy as a bedbug. Even among these vicious psychopaths, Benjamin Siegel stood out for his ruthlessness and violent temper. The larger-than-life mobster was one of the founding members of the notorious Murder, Incorporated, a squad of hit men dispatched to carry out gangland executions. A stylish dresser and charming womanizer, Siegel, with a full head of black hair and steely blue eyes, was as good-looking and wealthy as the movie stars he hobnobbed with when he moved his family to Hollywood from New York in the late 1930s and began muscling in on the prostitution, narcotics, and gambling rackets on the West Coast.

But he quickly realized the real money was in a place where gambling was legal—a tiny watering hole called Las Vegas, which had sprouted in the arid wilderness because of the presence of several underground springs. It wasn't much in the 1940s—a hellishly hot and dusty outpost situated at the intersection of three deserts, the Mojave, the Sonora, and the Great Basin. It was an empty no-man's-land between Utah and California with a train depot, a couple of dude ranches, and a handful of garish hotel-casinos that catered to tourists visiting Hoover Dam, 30 miles to the southeast. As World War II drew to a close, Siegel saw all the money being raked in by the cheesy casinos on the city's Strip. He had the uncanny foresight to see that gambling— legal and otherwise—could be an easily exploited gold mine in Las Vegas, even though Nevada itself had largely been settled by Mormons from neighboring Utah, who didn't indulge in games of chance or even alcohol. He convinced his gangland cronies to invest $1.5 million to construct a lavish

40-acre Havana-style resort with two swimming pools, a golf course, tennis and handball courts, and a luxurious casino and hotel filled with posh amenities and staff in tuxedos. He called it the Flamingo, named after his leggy mistress, the infamous Virginia Hill, and invited his Hollywood pals, like Clark Gable, Lana Turner, and Joan Crawford, to the hotel's glittery grand opening over Christmas in 1946. Initially, the Flamingo was an utter disaster that hemorrhaged money—plagued by bad weather that kept the celebrities away, construction delays, and horrific cost overruns that some attributed in part to Siegel and his girlfriend skimming off the top and squirreling the loot away in Swiss bank accounts. Six months after the hotel's discouraging start, it did finally become profitable—but the notorious 41-year-old hoodlum had stepped on too many toes and was killed execution-style in a hail of bullets one steamy June evening in the living room of his Beverly Hills home.

Siegel's murder quickly faded from the headlines, but like so many celebrities who lived fast, died young, and left a good-looking corpse, his legacy lives on. Improbably, Siegel's dream has become a reality beyond his wildest hallucinatory imaginings. In the half century since, the once sparsely populated cowboy town has morphed into a world-class tourist destination with palatial resorts and more than 2 million residents who host 40 million visitors each year.

While Las Vegas's present incarnation owes much to a cold-blooded if enterprising killer, what made this austere desert landscape bloom was the seemingly endless flow of water—enough to fill the swimming pools, irrigate the golf courses, and fuel its runaway growth. In the first half of the 20th century, a series of great dams, aqueducts, reservoirs, and extensive irrigation canals constructed to tame the Colorado River, whose flow emanates from the snowmelt of the Rocky Mountains, made it the primary freshwater source for more than 30 million Americans in the Southwest today. These massive engineering projects—which included the Hoover and Glen Canyon Dams and the creation of vast reservoirs like Lake Powell and Lake Mead—were designed to even out the water supply available from the erratic and unpredictable weather patterns that left farmers with soggy, flooded fields in rainy years and desiccated ones during the frequent dry spells and to provide a reliable supply to urban areas. They controlled floods, stored excess water in wet years for release during the inevitable droughts, and produced enormous amounts of hydroelectric power. This system made it possible to cultivate crops in areas that normally were too dry to farm and established a lifeline for arid western cities like Los Angeles, Phoenix, and Las Vegas that enabled them to survive in what were once inhospitable wastelands—a "Cadillac desert," to borrow the title of Marc Reisner's landmark history of the battles to control water that led to the development of the American West in the 20th century.

Today, 90 percent of Las Vegas's water comes from the Colorado, but the city's epic growth is rapidly outstripping its available supply. The underground springs that gave Las Vegas its name—which in Spanish means "the meadows"—were long ago pumped dry. But the other equally critical reason is that outmoded water agreements that were struck nearly 100 years ago, before the development of the Southwest, are still in effect. Nevada was last in line when water rights to the 1,450-mile river were divvied up between the seven states sipping from the Colorado—California, Arizona, Nevada, Colorado, Wyoming, Utah, and New Mexico. Legend has it that the alkaline Nevada soils were so obviously unconducive to farming that the state's delegate wasn't even included in the negotiations for the historic 1922 Colorado River Compact. Nevada's share is a paltry 300,000 acre-feet a year, a tiny fraction of California's 4.4 million cut or even Arizona's 2.8 million allotment. An acre-foot is approximately 326,000 gallons, which is enough to supply two average families for a year. Nevada's cut was based on its urban needs—which were pretty much nonexistent at that time—and no allocations were made for agriculture, which was then the driver for the mostly agrarian West. Consequently, there is no slack in the system—if metropolitan supplies dwindle, the cities can't temporarily siphon off more from their rural cousins, which is something other states have done in emergencies.

What's worse for the entire Southwest is that these water shares were apportioned based on the expectation the Colorado would provide 17 million acre-feet of water every year. But as it turned out, the decades in the early part of the 20th century were exceptionally wet ones. Historical and tree-ring data going back hundreds of years suggest the average flow is closer to 13.5 million acre-feet. In other words, the Colorado—which has been called "the most regulated river in the world"—is overcommitted. As a consequence, every single gallon that flows down the river from the melting snow of the Rockies is already spoken for. Yet the system lurched along for decades with relatively minor squabbles because somehow there was enough surplus water stockpiled in the reservoirs to tide the region over in dry times.

No longer. Since 2000, the American Southwest has been in a protracted dry spell, which has shriveled water levels in lakes and rivers and slashed mountain snowmelt. But what's happened in the past decade may not even be the worst-case scenario. A recent analysis of tree-ring data showed clear evidence of an epochal Southwestern drought lasting more than 60 years a millennium ago. Hydrology experts worry that we could be blindsided by an unlucky convergence: a once-in-200-years megadrought caused by natural climatic variations happening simultaneously with serious dry spells induced by global warming. And at a certain point, these conditions will no longer be considered a drought because the climate will have simply become drier; indeed, climate models project that by the end of this century, the Southwest will be more like the Sahara. Even if rainfall doesn't diminish—which is highly unlikely—the warmer temperatures will suck up more water. The runoff from mountain snowpack will dwindle, demand will swell because the crops will need more water, and far more water from the vast reservoirs will be lost to evaporation. "We have this wonderful infrastructure where we plan for three- to four- or even five-year droughts, which was not a bad time horizon for the 20th century," said UCLA's Glen McDonald. "But that might not be what's in store for the future. The insidious thing about planning for episodic climatic events like a drought is: How do you really know where you are in the cycle? A 14-year drought will become more like what we need to plan for."

There is even a fifty-fifty chance Lake Mead will run dry by the end of this decade, according to a controversial report by researchers at the Scripps Institution of Oceanography at UC San Diego. They concluded that the way we use water from the Colorado River is no longer sustainable. Scientists looked at a combination of factors: past water usage, scheduled water allocations, population growth, and global warming changes—evaporation and diminished snowpack runoff are projected to reduce available water by up to 30 percent—and were astonished to find that the river system was running a net deficit of one million acre-feet of water annually. In other words, we're sucking that much more out of the river system every year than is being replenished by the normal hydrological cycle. If this trend continues, simple mathematics shows that the approximately 110-mile-long reservoir—the largest in the United States—could soon be drained. And if the lake drops below a threshold of 1,050 feet, it could interfere with the Hoover Dam's hydroelectricity generation, because the water level might be too low to drive the dam's turbines. But the even greater worry is that the water could soon drop below the level of the intake pipes, the giant funnels that shunt water out to the Las Vegas Valley and to the aqueducts that carry water to California, Arizona, and Nevada, depriving millions of drinkable water. By 2050, the warming planet may reduce the runoff flowing to the Colorado River by up to 20 percent. "We were stunned at the magnitude of the problem and how fast it was coming at us," study coauthor Tim Barnett, a marine physicist, told *Newsweek*. "The signals have been there for at least 20 years: less snow, warmer temperature in the mountains, less river runoff. The evidence has been here for some time; we just pulled all the data together. Make no mistake, this water problem is not a scientific abstraction, but rather one that will impact each and every one of us that lives in the Southwest."

With the Colorado potentially running dry and no surpluses to draw upon as a buffer, Las Vegas has faced some hard choices. Enter Pat Mulroy, who some say has the toughest job in water management: slaking the thirst of the mushrooming desert city at a time of escalating scarcity. As the general manager of the Southern Nevada Water Authority, which serves the city of Las Vegas and the surrounding suburbs, Mulroy is a clear-eyed pragmatist. Her blunt, no-nonsense style has made her something of a legend in the water world, which is mostly populated with crusty bureaucrats and engineers with pocket protectors, not immaculately tailored women with master's degrees in German literature. "She's my hero—she's like Clint Eastwood," said Gerald Galloway of the University of Maryland, and only half in jest. But the suite of strategies that Mulroy has bulldozed Las Vegas into adopting over her two-decade-plus tenure may well serve as a model for other cities to implement.

Local folklore has it that within her first year on the job, she convinced casino titans like Steve Wynn, head honcho of the Mirage, Treasure Island, and Wynn, to stop their profligate use of freshwater in their extravagant water features—lakes, tropical lagoons, fountains, and lush golf courses—and substitute recycled wastewater. She even got them to pony up thousands of dollars to pay for the transition. She instructed her staff to scour arcane water bulletins issued by state agencies to figure out how much unclaimed water there was in the rural areas, in underground aquifers and springs. Then they filed applications for the rights to nearly a million acre-feet of unclaimed water in basins underneath four counties—more than three times the amount of water they are allowed to siphon from the Colorado each year—in a desperate search for this precious resource in case the river runs dry. "You cannot conserve 90 percent of your water supply, it is physically impossible," Mulroy told me in a recent interview; that's how much of their water they'd lose if the Colorado dries up. "We have to have a backup supply to reduce our reliance on the Colorado."

Until the recession hit, Las Vegas was flush with cash—thanks to all those tourists—and Mulroy used it liberally to exploit every angle that would keep the taps flowing. The city even explored the possibility of building a desalination plant on the Pacific Coast and exchanging that water for a share of California's rights to the Colorado—an idea they've abandoned for now because desalination is still too expensive and too energy-intensive to be a workable solution here in the United States. Mulroy executed complicated water-banking arrangements with California and Arizona and anted up $350 million for dibs on Arizona's unused surpluses in case of emergencies. They've also bankrolled ambitious brick-and-mortar projects to stay afloat: They're tunneling underneath Lake Mead to install another intake valve that would work until the water level dropped below 890 feet, and hope to soon break ground on a controversial 285-mile pipeline to transport groundwater to Las Vegas from ancient aquifers under the Snake Valley, on the Nevada–Utah border.

Las Vegas has also adopted some of the most stringent water conservation rules in the nation, which has enabled the city to shrivel its use by about one-third over the past decade even as the population has spiked by 400,000. They identified ways to ruthlessly cut usage without affecting quality of life.

In 1999, the city began its "cash for grass" program, paying residents an average of $1.50 per square foot to pull out their lawns and replace them with native vegetation, which has thus far saved the city more than 10 billion gallons of water (although outdoor use for lawns and car washing still accounts for 70 percent of the local water use). If Mulroy had her way, one reporter wryly noted, she'd rip out every blade of grass in the county. Starting in 2003, the city began relentlessly prodding the public to use water more efficiently, including restricting watering days, fixing leaky spigots, and hitting people with fines if they wasted water, to wring out every excess drop. They now recycle virtually all water used indoors by capturing, treating, and reusing it, and slashed their per capita water footprint to 75 gallons a day (in comparison, residents of Tucson use 114 gallons, while Angelenos consume 125 gallons).

"People who move here have to recognize this is the driest desert in the United States and we average four inches of rainfall a year," Mulroy said. "We brought with us magnolias from Florida and Kentucky bluegrass— things that require an extraordinary amount of irrigation to stay alive. We can't do that anymore. We became really aggressive eight or nine years ago, which gave people time to adapt and do things in a systematic way that hasn't disrupted the economy. But you have to slowly make permanent changes—you can't expect an entire community to turn on a dime or to take draconian measures that can't last forever." Unless, of course, you're Clint Eastwood.

CHAPTER SIX

THROUGH A GLASS DARKLY

Jordan Rice was an unlikely hero. Barely 13, with tousled strawberry-blond hair and inquisitive blue eyes, he was a shy, quiet eighth-grader in a comfortably middle-class family. But the teenager's bravery on a warm summer day in 2011, in the midst of possibly the worst floods in Australian history, transformed him into a national symbol of courage and hope for the victims of the disaster. He lived with his parents and three brothers in Toowoomba, Queensland, a farming community about 80 miles inland from Australia's east coast that sits at the crest of the Great Dividing Range, the string of mountains that separates the coastal lowlands from the rich soil and abundant grasslands of the agricultural heartland.

I spent nearly a month in Australia a few months after the floods, traveling all over the country to see firsthand the effects of climate change in an advanced, industrialized democracy. When I drove through Toowoomba, a community of about 120,000 people clustered around low-slung buildings, quaint shops, and family-owned businesses, I was struck by its resemblance to the Midwest of the early 1960s. In good years, Toowoomba was clearly a prosperous town, a hub of commerce in the fertile farming and grazing region known as Darling Downs, where much of the nation's food and livestock are produced. Tourists visit to see the well-manicured English-style gardens that dot the city.

But I could also see the region had fallen on hard times after a brutal, nearly decade-long drought that started in 2003—perhaps the worst in 1,000 years—turned the verdant fields into parched dust bowls. Harvests of wheat, barley, and canola dropped by 60 percent, hundreds of acres were left

141

fallow, storefronts in Toowoomba's once-thriving downtown were boarded up, and more than 10,000 families who had farmed for generations in the surrounding agricultural areas were forced into bankruptcy or simply abandoned their land because they couldn't find buyers. Severe water restrictions had been in place for years and the city, in an act of desperation, hoped a state-sponsored cloud-seeding experiment would end the long dry spell.

Yet when the rains finally did come, they did nothing to ease the drought. Months of torrential spring storms that began in November 2010 and continued through the Australian summer and into the new year drenched vast stretches of the northeastern state of Queensland and created an inland sea the size of France and Germany combined. But instead of replenishing groundwater, the relentless showers washed away the fertile topsoil and turned grape-growing vineyards, orchards once covered with neat rows of fruit trees, and fields of cotton plants into soggy marshes.

There had been a welcome break in the weather, and it was hardly raining that afternoon in January when Donna Rice, Jordan's 43-year-old mother, decided to venture out with Jordan and his 10-year-old brother, Blake, to pick up one of their older siblings after school. Unexpectedly, the skies opened up and a cloudburst from a freak storm sent a raging torrent of water through the Lockyer Valley, a natural trench carved between the mountains just east of the city, unleashing a tsunami-like wall of water that roared through the streets at lightning speed, demolishing everything in its path.

Engulfed by the rising floodwaters, Rice's white Mercedes stalled at a busy intersection. Frantic, she called the emergency dispatch, which was overwhelmed with other calls. When the operator finally came on the line, she instructed Rice to climb onto the roof of her car with the two boys and wait. Horrified bystanders watched helplessly as the water quickly rose around the trapped mother and her two young sons until a Good Samaritan truck driver arrived on the scene, tied a rope around his waist and jumped into the raging torrent. He reached for the older boy, but Jordan refused to go first, even though he couldn't swim and was terrified of the deep water.

"Save me brother," he pleaded.

Blake was whisked to safety, but before the driver could return to rescue Jordan and Donna, they were swept away by the floodwaters and drowned.

"I saw the panic in their eyes," one witness who tried to help with the rescue but was knocked off his feet later told reporters. Their bodies were found an hour later more than a mile away. Hundreds of mourners attended their funeral, many wearing powder-blue ribbons in memory of the teenager's sacrifice, which became a touchstone for countless communities devastated by floods that grew ever fiercer.

The day after Jordan died, the rains moved eastward, filling the already swollen Brisbane River, which snakes through the city of Brisbane. The sophisticated coastal metropolis, home to more than two million, was flooded when the river overflowed its banks. A 10-foot-high surge of water gushed as if from a burst dam through the pocket parks and tree-lined streets of the city's downtown, saturating the gourmet restaurants, trendy boutiques, upscale hotels, stately government buildings, and gleaming glass skyscrapers that lined the riverbanks with a thick layer of mud, sediment, and debris. Thousands of people in low-lying areas who were directly in the path of the rushing waters were ordered to evacuate their homes and businesses. Supermarket shelves were emptied by thousands of panicked shoppers snapping up staples. The only open highway was gridlocked with residents fleeing in a frantic mass exodus.

By Wednesday, the central business district was a ghost town. Black Hawk and Chinook helicopters filled the skies, transporting supplies and rescuing stranded residents from rooftops. Lines for sandbags stretched more than a mile from relief centers, which were filled to capacity with people left homeless. There were rolling electricity outages all over the city, and in some submerged suburbs only the crowns of trees remained visible. The airport at the northern end of Brisbane, an ultramodern steel-and-glass hub for flights throughout Australia and the Pacific, was shuttered because the runways were underwater, forcing troops to fly into nearby towns and truck fresh food, water, and emergency supplies to a city the size of Houston. "It was unreal," Brisbane resident Narelle Sutherland told me when we had lunch at a posh bistro on the rebuilt waterfront a few months later. "I've seen these kinds of things on the news, but I never expected to be caught in the middle of it in my own hometown."

Yet less than three weeks later, in February, Cyclone Yasi, a monster Category 5 storm with a 300-mile front and core winds of up to 180 miles an hour, roared in from the Coral Sea and slammed the northeastern coast, flattening several beachfront towns, many of which were still cleaning up from the floods. The largest cyclone to hit Queensland in nearly a century, Yasi left hundreds of families homeless. More than 30,000 were evacuated from Cairns, a sultry tropical paradise of 160,000 that is the jumping-off point for dive expeditions to the Great Barrier Reef. Think Miami Beach in the 1950s without the gangsters and the movie stars. Parts of the city were without power for five days, hospitals were emptied, and 100 patients were airlifted 1,000 miles south to Brisbane.

Ultimately, the storms claimed 53 lives, left more than 100,000 people homeless, and forced thousands more to evacuate. Vital food crops were destroyed, pushing prices of staples like bananas to $7 a pound; wine growers' crops were decimated; hundreds of cattle were drowned and washed away; and the lucrative coal-mining industry was brought to standstill. There was more than $20 billion in damage overall. Dozens of villages along the coast, along with the banana and sugar plantations in the northern tropics, were left in utter shambles, looking like a pack of untamed three-year-olds had gone on a rampage with Tinkertoys.

These calamities are among the most recent snapshots from a continent under siege. The Land Down Under is on the front line of climate change, experts say, the proverbial canary in the coal mine that exemplifies what global warming models have long predicted. "Our recent extremes are quite strongly climate-change driven," said Penny Whetton, a climatologist at the Commonwealth Scientific and Industrial Research Organisation (CSIRO). "And they're an illustration of what we expect to see more of in the future, when natural fluctuations are intensified by global warming."

The wild weather systems that have plagued the United States in the last few years and given us a whole new vocabulary to describe once rare phenomena—the severe windstorms known as derechos and the wind-whipped walls of dust called haboobs that swept across the Plains and the Southwest—have gripped Australia for more than a decade. If you want a preview of what life will be like under unrestrained global warming, these experts darkly warn, spend time in Australia, one of the nations most vulnerable to climate change. But that's also why some of the best solutions for coping with the destructive forces unleashed by a warming world are coming from the hardy island continent.

THE DESERT CONTINENT

Australia is flush with natural resources—uranium, coal, oil, gold, zinc, nickel, and the rare earth minerals used in cell phones and electronics—and blessed with sparkling, pristine beaches that extend for thousands of miles, as well as with the world's oldest rain forests, lush tropical jungles with dense canopies that cover the northeastern tip of the continent. It's a big country, almost as large as the continental United States, but with fewer people—about 22 million—than live in the state of Texas. Aside from the tropical belt in the north and a fringe of green along the coasts, Australia is one big desert island. Half the population is clustered in three major cities that hug the eastern coast: Sydney, Melbourne, and Brisbane.

Its geographic and social isolation have insulated Australia from the wars and economic upheavals that fractured other developed countries. Until the 1960s and the advent of affordable international air travel, it was a remote colonial outpost of the British Empire, a thriving democracy surrounded mainly by impoverished Third World dictatorships. The country was settled mostly by outcasts looking for better lives—first as a penal colony and later by waves of immigrants from southern Europe, primarily Italy and Greece, which created a nation infused with a hardy, pioneering spirit that enabled Aussies to carve out an enviable lifestyle in some of the world's harshest terrain.

This may explain why Australians, despite differences of class and political outlook, retain an almost small-town sense of shared national purpose. Sure, there are progressives and conservatives and every stripe in between, but the toxic polarities that have split America into warring red and blue factions are nonexistent there. As Thomas Friedman wryly observed, just about their entire political spectrum would fit comfortably inside our Democratic Party. And when people die in natural disasters, each loss is felt keenly. A remarkably civilized country, there is a strong belief in the importance of a social safety net for the less fortunate, and the minimum wage is $15 an hour—

roughly double what it is in the United States—which assures everyone at least a decent standard of living. Universal health insurance guarantees access to medical care as a right, not a privilege, and doctors are salaried as part of an integrated health care system. Voting is compulsory—you get fined if you fail to cast your ballot—and the country is small enough that people feel an intimate connection with their leaders. "We're more open to government intervention for the greater good," said Elizabeth Hanna, a scientist with the Climate Change Adaptation Research Network for Human Health at Australian National University in Canberra. "This is what we expect. That's why we're quite happy to pay taxes and help those people who need it more."

But the laid-back Australian way of life may soon become a relic of history. The country has been long been whipsawed by extreme swings in weather and a changing climate has intensified the natural disasters that have swept across the continent and transformed serious floods and harsh droughts into catastrophes of biblical proportions; unleashed killer heat waves, agricultural collapse, bushfires of unimaginable ferocity; and hastened species extinction. These events are expected to become even more frequent and severe, according to a recent study by the Australian government on the risks of climate change, and could transform Australia in ways that are unimaginable today: killing off the Great Barrier Reef, drastically shrinking water and food supplies, swamping coastal cities like Sydney with the rising sea that will obliterate their iconic beaches and waterfronts, and rendering vast swaths of the country too hot and too dry to live in. Annual rainfall in already parched southern Australia is expected to decline by 50 percent.

For those who still cling to the notion that our nation's relative wealth will insulate us from the harshest effects of climate change and that it will impact mainly future generations of the poor in developing countries, Australia is a prime example that it is simply wishful thinking. In the not too distant future, this prosperous, advanced democracy could be completely destabilized: rent internally by internecine battles over increasingly scarce supplies of food, water, and other natural resources; overrun by desperate northern neighbors in Indonesia and the Philippines, who will have to flee the overheated tropical regions that are disappearing underwater; and crushed by the economic fallout of paying billions for reconstruction and adaptation schemes.

Within just the next decade, Australia is bracing itself for a host of health woes triggered by rising temperatures. These include a doubling or tripling of heat-related deaths in the elderly, even in major cities; higher rates of injury, death, and post-traumatic stress disorder in the aftermath of increasingly ferocious floods, storms, cyclones, and monsoons; changes in the range and seasonality of outbreaks of mosquito-borne infections such as dengue hemorrhagic fever, Ross River fever, Barmah Forest virus infection, and even malaria, which could become endemic as far south as Sydney; freshwater shortages, with consequences for hygiene and sanitation; and regional increases in the release of pollens and spores that can cause or exacerbate asthma and other respiratory conditions. Severe droughts and long-term drying conditions pose their own health-related problems such as depression and suicide, developmental and emotional problems in children, and sickness resulting from exposure to extremes of heat, dust, and smoke.

During the latest drought, which lasted from 2003 until the recent floods, the collapse of local economies increased threats to mental health and sparked negative changes in health-related behaviors, including domestic violence, alcoholism, drug addiction, and smoking, numerous studies have shown. Families unraveled under the corrosive strains and social cohesion broke down because of the loss of jobs, homes, and comforting routines. After being crushed by bad weather, one in five people experienced "extreme stress, emotional injury and despair," according to a recent study by Australia's Climate Institute, which noted "this is a 'new normal,' for which the past provides little guidance."

The future doesn't look much brighter. Within the next 20 years, said James Smith, a health official in Australia's Northern Territories, vast portions of the northernmost regions of the country could be submerged under rising seas. By midcentury, if global CO_2 emissions continue unabated, the frequency of exceptional droughts in the south, like the one that crippled the nation for nearly a decade, is likely to increase from one every 25 years to one every 2 years. In the coastal city of Darwin, a multicultural urban center of 130,000 on the northernmost tip of the continent and the gateway to Indonesia and East Timor, temperatures now hover near 100°F on 20 to 30 days annually, and the effects of the heat are magnified by the high tropical humidity. In the next few decades, scientists predict Darwin will melt under record-breaking heat on 300 days a year, making it virtually impossible to live there. "We're looking at temperatures in excess of 120 degrees," said Dr. Peter Tait, a physician in Alice Springs in the Northern Territories and a prominent climate change activist. "There is a ceiling of temperature beyond which no amount of acclimatization would allow us to efficiently cool."

By the year 2100, if even the midrange prediction of a sea-level rise of 2 feet—and that's a conservative projection—occurs, 100-year flooding events will occur several times a year throughout the country and 10-year floods will happen every 10 days. Only a relatively thin band of land around the edges of the desert continent is habitable now. As climate change eats up more arable land, with even drier weather and rising sea levels claiming big patches of the coastline, Australians could soon run out of livable space. "These rapid changes start unraveling the fabric of the ecosystem and adaptation can only go so fast," Tait, who has the full beard of a rabbinical scholar, told me when we met for dinner in Cairns. "If change is happening faster than you can keep up, there will be strife until you hit the new equilibrium. That's what's frightening—the level of risk is increasing at the same time that our capacity to adapt is being reduced."

If temperatures climb by 8°F, as is predicted for Australia by the end of this century, it "would be unlike anything experienced before by modern human societies, presenting us with huge challenges in terms of our ability to adapt," said CSIRO's Penny Whetton. "We would be on a different planet, and not the planet we evolved in," added Peter Tait. "We're talking environmental meltdown."

In other words, as Elizabeth Hanna succinctly put it, "Which part of 'the house is on fire' are you not getting?"

AFTER THE DELUGE

Following the recent floods, the stoic Aussies did what they always do after being walloped by the weather, and simply hunkered down. Australian Army Major General Mick Slater, a lanky, ramrod-straight, no-nonsense military man who graduated from the US Army War College in Pennsylvania, was put in charge of the recovery effort. Crews worked 24-7 for months, repairing roads and cleaning up the detritus of the flood—broken furniture, demolished homes and buildings, overturned cars, and other debris that littered the fields and streets. In Brisbane, battalions of strangers showed up unbidden on people's doorsteps—"a mud army," said Narelle Sutherland—and spent long hours knee-deep in wreckage helping their less fortunate neighbors clean up.

Like an embattled prizefighter, Oz is punch-drunk from its escalating war with nature, and scars from these skirmishes are tattooed across the island continent. The nation's capital, Canberra, is a parklike planned city of 345,000 on a patch of land about halfway between Sydney and Melbourne, an upscale company town of mostly highly educated bureaucrats. Wide, tree-lined boulevards radiate from Lake Burley Griffin, an artificial lake that's surrounded by ornate, magisterial buildings that house government agencies, the Parliament, cultural institutions and museums, and Australia's world-class national university. The bare hills and gentle valleys of the Monaro Plains outside of Canberra were once home to herds of sheep that grazed year-round on the rich native grasses.

But when a pilot buddy of Elizabeth Hanna's obligingly gave me an aerial tour of the city in his four-seater Cessna, I could see that huge swaths of the surrounding countryside were still blackened by a firestorm that threatened to engulf the city a decade ago. After months of scorching weather, the land was so dry that the pine forests turned into tinderboxes. About 30 miles southeast of Canberra is the Googong Reservoir, a long finger of water that stretches over 650 acres when filled to the brim. Over a million people drink from the reservoir, but the recent drought shriveled it to less than 30 percent of its capacity. Even after the drenching rains, water levels remain dangerously low.

Hanna has a hard time reconciling another shrunken body of water, Lake George, with the shimmering bowl of her youth. "When I was a child, we'd drive up from Melbourne and it would be overflowing its banks," she told me one sunny Saturday afternoon in April, shaking her head in disbelief as we stared across the scrub- and silt-covered dunes now filling land that was once submerged. "And now look at it."

Meanwhile, in the northern tropics, warming waters in the Coral Sea and the Gulf of Carpentaria are unleashing ever more powerful cyclones, scorching heat waves, and longer and wetter rainy seasons. And then, of course, there are the floods, like something out of Genesis when it rained for 40 days and nights and Noah set sail in the ark—except in Australia, the storms went on virtually unabated for more than two months. Yet much of the country was gripped by a fierce drought in the first decade of this century. Draconian limits on water usage were in place throughout much of the southern and eastern parts of Australia. Dinner table conversations revolved around the best places to score rainwater tanks and water-efficient showerheads. Evening news shows regularly announced water levels in the nation's reservoirs and featured stories about scofflaws being slapped with heavy fines after neighbors caught them hosing down their cars. In one celebrated incident, South Australia's water minister quenched his own thirsty lawn in a flagrant defiance of the rules, provoking a public outcry that led to his resignation.

In 2009, the Southern Ocean coast city of Adelaide, which is the driest city on the world's driest inhabited continent, came perilously close to running out of drinking water, and city fathers were literally days away from trucking in bottled water for some of the area's million-plus residents. Farmlands in the once fertile Murray-Darling Basin, an area the size of France and Spain combined where Australia's three main rivers converge, turned into dry rubble. Power plants were shut down for lack of cooling water. Harsh dust storms rolled across the bottom half of the country. Even Sydney was enveloped in the worst haboob in its history in September 2009, covering the city in a thick shroud of grit that turned day into night.

Earlier that year, the parched terrain had provided kindling for raging firestorms that incinerated entire towns. A suffocating heat wave had smothered much of southern Australia in temperatures of more than 100°F for nearly a week, making it the worst hot spell in a century. Rail lines buckled, more than 140,000 homes were without electricity when the power grid melted, bats were dying in droves, and city dwellers were leaving pans of water at roadsides for wilted kangaroos. More than 700 people were hospitalized and nearly 400 died. "It was a silent disaster in a way, because so many people died at home," said Dr. Marion Carey, program leader for the Monash University Climate, Biodiversity and Health Program. "The population isn't acclimatized. You get these sudden huge spikes in temperatures and people can't deal with it. We had three consecutive days with temperatures above 110 degrees. Emergency calls for ambulances jumped and the number of patients arriving dead on arrival tripled. The morgues were filled to capacity—and this was all before the bushfires started."

The unprecedented heat coupled with the long dry spell turned hundreds of thousands of acres of farmland into a giant matchbox that ignited a firestorm of unimaginable ferocity across a huge chunk of southern Australia. Because of the extreme temperatures, the ground was baked and there was no moisture in the soil. Fires have ravaged Australia for thousands of years, but the one that exploded into flames one afternoon in February 2009—a day they now call Black Saturday—was different because the fires were exceptionally hot and came from every direction, a phenomenon known in firefighting circles as spotting. Smoke, ash, and burning embers from the bark of trees were carried aloft for up to 20 miles by wind and updrafts in the fire's convection currents, jumping across roads and clearing firebreaks to start new fires on the parched ground. These separate fire fronts coalesced into the walls of the impenetrable infernos that trapped so many people. Huge plumes of flame with temperatures as high as 5,000°F shot hundreds of feet into the air, setting ablaze even wet mountain ash forests that are normally immune to fire. "It sounded like a jet engine roaring overhead," Carey recalled.

At least 173 people died in the flames, the highest death toll in Australian history; thousands were left homeless and a quarter of a million acres of land were torched. Most of those who perished didn't stand a chance simply because there was no way out. Some died getting out of their cars in futile attempts to outrun the fires or find shelter. Others tried to return to their homes and farms to rescue their animals and died with them. Across rural Victoria, police and firefighters discovered dozens of vehicles containing bodies charred beyond recognition. In some cases, entire communities were ringed by flames and were unable to escape to safety. "Marysville got wiped off the map," said Carey. "One minute a town is there and the next it is not. How do you deal with that? Everyone recognized there was a quantum shifting in the nature of the bushfires. We've always had heat and fires, but the extraordinary severity—that was new."

Even those who weren't directly in the path of the fires will be forever haunted by memories of the inferno. "We had suffered through two weeks of extraordinarily high temperatures," recalled Corey Watts, a regional projects manager at the Climate Institute in Melbourne, near the fire zone. "It was up to 116 degrees Fahrenheit during the day—a record for Melbourne—and hovered around 104 degrees at 10 o'clock at night. The next thing you know, there was a huge cloud over the city from the fires and choppers were crossing the city bearing the wounded in and out, trying to get water to dump on the fires."

Watts told me a friend who had stayed behind to save his house was hosing it down when he was engulfed by flames. "He was on the phone with another friend who said you could hear the roaring fire in the background," Watts recalled. "Suddenly, he said, 'I've got to go,' and he took off in a car with his wife and baby daughter in a desperate attempt to outrun the firestorm.

"They didn't make it."

THE DRUMBEAT OF CATASTROPHE

Although houses can be rebuilt and highways repaired, the extensive monetary hits from the weather-related calamities are eroding the fiscal health of a country that largely avoided the excesses and therefore the subsequent meltdown of the worldwide financial crisis. But the steady erosion of health and well-being triggered by rising temperatures, along with the attendant environmental degradation, aren't so easy to fix. And the constant drumbeat of catastrophes exacts a heavy psychological toll—whether it is prolonged and insidious from a drought or brutish and short, like the bushfires—especially for the thousands whose lives are upended, leaving them homeless, stuck living in cramped trailers or motels or with their extended families for months or perhaps years. Even the famously resilient Aussies have been worn down.

Like Peter Goodwin. "The river came up to right where we're sitting and the waters were more than two feet deep," he told me while we sat in the driveway of his ranch-style house perched on the banks of the Balonne River in Saint George, a picturesque village of 3,500 in eastern Australia's wine country. It was a drizzly Sunday afternoon in April, three months after the devastating flood of 2011, which had come right on the heels of a smaller flood that had washed away the town nine months earlier, in March 2010. "These floods were once-in-150-year events and we had two of them in less than a 1-year period," said Goodwin, 60, a crusty salt-of-the-earth type with alert blue eyes and callused hands. He works as an operations manager overseeing the upkeep of parks and gardens for the local municipality—or shires, as they're called there. "People who lived here never had seen floods this high."

During the most recent storm, the rain-swollen river rose to a record 45 feet, cutting off the town from the outside world. The flat, dusty plains of agricultural areas like Saint George were hardest hit by the drenching downpours and the overflowing rivers that made roads impassable, cut off power lines, washed away vineyards and fruit orchards, drowned hundreds of cattle, and covered homes and everything inside them in thick layers of sediment and mud. Shell-shocked residents like Goodwin were still burrowing out from under the debris.

"That's the hard part of the flood—the aftermath," said Goodwin, who was forced to move in with his grown daughter twice while he made his home livable again. "You get a lot of help during the flood, but then everyone settles back into their routine. There are a lot of houses down there that are still empty," he added, shaking his head as he gestured toward the riverbank. "And they will be for a long time to come."

Saint George is a throwback to a different era, a place where everyone knows each other and half the town has dinner on Saturday nights at the local R&S, a civic organization similar to the Elks. The village was marooned by floodwaters for nearly two weeks, but city and public health officials had a well-coordinated disaster management plan in place. Because it's so small and they had time to prepare, they had stockpiled medication, and at the outskirts of town they'd set up a makeshift field hospital in a specially outfitted tent that can accommodate 15 beds. Seven rescue helicopters airlifted seriously ill patients to hospitals in Brisbane while two others remained on standby in case of emergencies.

But in many towns like Saint George, services were stretched to the limit because of the magnitude of the disaster. Everyone pitched in, of course, but eventually they were ground down by the sheer futility of fighting their losing battle with the forces of nature, of working feverishly around the clock to create four-foot-high barricades of sandbags around their homes and businesses to stave off the looming floodwaters. "We had about 100 volunteers from around town with earth-moving equipment," recalled Dr. John Hall of Oakey, a creek-side town of 3,600 about a dozen miles outside Toowoomba. "But everything got washed away and their businesses got inundated anyway. We had 130 houses go under, which is unprecedented in this town. Afterwards, hundreds of families were homeless and we still have people living in motels months later. A lot of them are struggling financially. Some were uninsured and have to start again from the ground up. And as disasters become more frequent, there is the fatigue issue, and generosity dries up."

It's the same story everywhere else, "a pervasive sense of helplessness and hopelessness," Dr. Dan Halliday said as he showed me around Stanthorpe, a picturesque Australian wine-country village of brick cottages and craftsman-style homes that was inundated in the recent flood. A strapping six-foot-four-inch Hugh Jackman look-alike, the 30-something Halliday at the time was president of the Rural Doctors Association of Queensland. He comes from a farm family and feels a deep obligation to serve the farming community, which has hit upon hard times. "Farming has always been difficult and my parents encouraged me to do something else with my life," he recalled when we stopped for a lunch of tea and shepherd's pie at a local café. "'You've got brains,' they'd say. 'Do something for yourself. And if you want to come back to the country, then it's on your terms.'"

He became a doctor and did come back with his wife, whom he had met as an undergrad. But they essentially walked into a decade-long drought, and along with it, an overwhelming sense of grief that a way of life that had sustained thousands of Australians for nearly two centuries was drawing to a close.

Australia's farmers don't have the safety net of government subsidies like we do here in the United States, yet agribusiness provides nearly two million jobs, more than 20 percent of the nation's exports, and the vast majority of the food Australians consume. But droughts are the new normal, which triggered a precipitous drop in farm property values. At the height of the dry spell in rural Victoria in the south, suicide rates rose by nearly 10 percent. Debt-strapped ranchers and farmers took their own lives at the rate of one every four days during the long dry spell. Children were on suicide watch, remaining vigilant for signs a parent was underwater in despair, but suicides went up among the youth, too.

Within the next 50 years, Australia could be forced to import food to feed more people at precisely the time when supplies are scarcer—a possibility that is anathema to the proudly self-reliant Aussies. "There's a real question about our ability to provide food for ourselves down the track," Halliday grimly observed. "There was a lot of despair, particularly toward the end of the drought. Then farmers finally get to the stages where they can plant crops again, and then the floods hit. After a prolonged period of time suffering mental anguish and hardship, it's a big double whammy and caused a lot of people to consider suicide. The rural youth have access to more violent methods of suicide—very hard, very nasty, brutal methods—because of this level of despair. Very toxic chemicals they ingest. Firearms. They feel like they just can't cope anymore."

WHY AUSTRALIA?

Much of Australia's vulnerability results from how it evolved as a continent and its geographic location at the intersection of three oceans—the Indian, Pacific, and Southern. About 80 million years ago the Australian landmass broke away from Gondwana, a supercontinent that once encompassed most of the continents of the Southern Hemisphere, including South America, Africa, and Antarctica. Almost completely covered in rain forest at that time, Australia then drifted northward toward the equator in complete isolation for 30 million years, during which time the central regions of the country dried out and created the world's most ancient deserts, which are filled with unique indigenous trees and brush that can exist on tiny amounts of water.

Australian soils are deficient in vital nutrients such as nitrogen, phosphorus, and zinc, mainly because the region is old, even in geologic time, and most of the country hasn't been revitalized by the soil-renewing activity of volcanoes or glaciers in more than a million years. The fragile river systems have been exploited for irrigation, which has hastened soil degradation, made the ground saltier, and contributed to creeping desertification. "The country is already dry and ecologically fragile with relatively poor soils," said Steven Sherwood of the Climate Change Research Centre at the University of New South Wales. "It doesn't have the resilience, so it's more vulnerable to anything."

Some of the country's higher risks are simply logistical. Roughly 85 percent of the population is crammed into coastal cities, making them more vulnerable to sea-level rise, storm surges, and cyclones. But the prime driver of Australia's unique susceptibility is its location in the midst of three oceans. The island nation is buffeted by the complex interaction between their currents—the El Niño–Southern Oscillation in the south Pacific, the Indian Ocean Dipole, and the Southern Annular Mode—and their normal oscillations may be speeding up from the steadily warming waters. The Pacific has historically switched between El Niño and La Niña phases, triggering correspondingly dry and wet decades in Australia.

During the El Niño, warmer waters in the southeastern Pacific create air pressure systems that drive rain bound for eastern Australia out over the ocean. La Niña, in contrast, has cool ocean temperatures, which lead to stronger easterly winds in the tropics that pile up warm water in the western Pacific and around Australia. This causes more water to evaporate into the atmosphere, intensifying monsoon rains and cyclones. In fact, some of the strongest El Niño patterns ever recorded were blamed for the crippling drought, while an equally intense La Niña precipitated the downpours that inundated Queensland.

The Indian Ocean Dipole, which was discovered only in the past decade, cools the eastern tropics and warms western currents, reducing spring rains in southeast Australia. And changes in the Southern Annular Mode, a climate system that prevents rain-bearing low pressure systems from passing over southern Australia, has virtually stopped winter showers in Perth. This climate pattern has been steadily shifting southward toward the South Pole since the early 1970s, pushing rainfall out over the ocean instead of onto southwestern Australia. Rainfall has dropped by 20 percent, and at least half of that shortfall is due to heat-trapping greenhouse gases, research shows. "It's been quite dramatic and much stronger than the climate models have predicted," said Penny Whetton of CSIRO.

There are now worrisome hints that warming water is tunneling away underneath the ice shelf in Antarctica. "Breaking up of glaciers in Antarctica will cool the surface temperature of the Southern Ocean, which in turn could make wind patterns more intense, causing more severe and frequent cyclones," said Albert Gabric, an associate professor at the Griffith University School of Environment in Brisbane. "Australia has all these natural regional climate patterns, which makes it a very complicated place to study. There are multiple stressors, creating layers upon layers of complexity, and warming adds an extra dimension."

SAVING OZ

Because of its unique vulnerability, Australia has become a living laboratory for adaption to a warmer world. Its scientists have devised innovative strategies to better manage scarce water resources, halt the spread of infectious diseases, and squeeze more crops out of increasingly dry land. In fact, if necessity is the mother of invention, looming catastrophe is the father.

Throttled by a severe water crisis that threatened the country's very survival, the Australian government launched an unprecedented $10 billion water security campaign that is a model for what needs to be done when the tap is running dry: They radically transformed the way water is handled on every conceivable level, from allocations for agriculture to how much flows out of faucets in Sydney high-rises, and dramatically swelled supplies through more efficient use and turning the seas into potable water at desalination plants.

Their Water Resources Observation Network (WRON) is one cornerstone of that campaign. The high-tech water management program was devised at Canberra's Black Mountain Laboratories, a leafy, college campus–like facility spread over 350 hilly acres that is the main research center for CSIRO, Australia's national science agency. Stuart Minchin, an environmental chemist and

IT whiz who heads the WRON team, showed me around a climate-controlled room where hard drives the size of meat lockers with 500 terabytes of storage space—the entire Library of Congress eats up only 10 terabytes—quietly hummed, crunching vast amounts of data. All that processing power was harnessed to corral the information collected by the nation's diverse water management agencies into a single grid and track virtually every drop of water in lakes, reservoirs, aquifers, and river systems.

"If we can't measure it, we can't manage it," said Minchin, who looked like a hybrid of surfer dude and apparatchik in a snappy black pin-striped suit, an earring in his left lobe, and wraparound sunglasses perched atop his head. He oozed jaunty Aussie charm. But make no mistake: His team is building perhaps the world's most ambitious national water information and accounting system, one that will be the basis for drastic changes in how water is used and distributed.

Inside WRON's nerve center, Minchin strode past four flickering computer monitors perched throughout the darkened state-of-the-art computer lab. He flipped on a vividly colored graph that illuminated the arcane details of water management, such as average rainfalls and projected evaporation rates, on a jumbo projection screen. Contained in his simple presentation was more than three years of monumental and mind-numbing work culling information from the 600 different organizations across Australia that collect water data—including private industry; local, state, and federal governments; and a variety of water allocation authorities.

Minchin fed this data into computer models that, given different climate scenarios, could help him project future water availability, in order to make some hard choices about how to equitably divvy up this vital resource. The next step will be to embed low-cost wireless sensors in waterways to continuously feed data into WRON's computerized network, enabling the monitoring of water resources across the country in real time, flood management, and allocations of water for consumption, hydropower generation, and agricultural irrigation.

Australia is one of the world's breadbaskets, but throughout the 20th century, the government encouraged farming on land that couldn't sustain it and handed out water licenses to just about anyone who was willing to put in the backbreaking toil of making the scrub desert miraculously bloom with citrus orchards and vineyards. These conflicting uses placed increasing demands on an overburdened and poorly managed river system, and it pretty much collapsed when the drought hit. "We had successful farmers in desperate straits that were forced out of business because they couldn't get enough water to keep their trees alive," said Minchin. "That was our Katrina moment. We knew we had to develop a scientific basis for determining if these rights and entitlements are compatible with each other."

WRON is just one strand of a comprehensive water policy program that is expected to quench this dry continent's thirst throughout this century, even in a warming world. "You need a portfolio of approaches in managing water security," said James Cameron, an economist and CEO of the National Water Commission, the Canberra-based government agency in charge of Australia's water reform agenda. "The right mix will depend on what's available in different parts of the country."

The government has imposed severe limits on water consumption, and they are rigorously enforced with stiff penalties. Wasting water has become as socially unacceptable as smoking around kids, and people are encouraged to blow the whistle on errant neighbors who fail to follow rationing rules. Almost 90 percent of households are metered, and prices have been raised by up to 40 percent in some cities to reflect water's true cost, which has curbed use, too. Extensive public health campaigns educated residents on the best ways to conserve water, such as taking four-minute military-style showers, not running water in the sink, installing water-efficient taps and toilets, recycling rainwater and household water for gardens and lawns, and using wastewater for irrigation and industry. Rainwater tanks and water-efficient showerheads became mandatory across the country.

These draconian conservation measures, along with more efficient methods of water use, enabled cities to cut water consumption by up to 90 percent over the past decade despite population increases. In water conservation circles, this dramatic alteration of habits is like the creation of a virtual river, akin to suddenly discovering another Mississippi.

Industrial water use has also been reduced by more than 30 percent—and nothing is sacrosanct. Despite the clout of the farm lobby, large cutbacks in water allocations to agricultural producers in the Murray-Darling Basin have been proposed to restore the health of its river systems. At the height of the drought, the rivers' flows had dried up to 5 percent of their averages, giant red gum trees were dying along the banks, dead fish floated in deoxygenated pools, and the soil was baked bone-dry. The Australian government also earmarked more than $50 billion for water improvement measures, which include more efficient farm irrigation methods and constructing massive new water pipelines to prevent leakage.

But the biggest source of water may ultimately be drawn from the vast oceans surrounding Australia. In 2006, the first reverse osmosis seawater desalination plant opened in Perth, and it now supplies 17 percent of that city's drinking water. Desalination is a process that removes salt from seawater, creating freshwater that is good enough to drink. Most plants—currently there are three, with three more about to come online and another in the planning stage—use reverse osmosis, in which seawater is pumped at high pressure through an ultrafine membrane that filters out the larger molecules, including salt. It's an expensive and energy-intensive process because salt binds tightly with water chemically, making it difficult to separate these molecules. But desalination plants that are now operational or under construction in Brisbane, Melbourne, Adelaide, and Sydney will yield millions of gallons of potable water that could eventually supply up to 30 percent of the water for these cities. These extensive measures are only the beginning, and researchers are busily looking at cheaper, more energy-efficient ways to extract water from the sea and even better methods of conservation in urban areas and on the farms that suck up most of the available H_2O. "The recent rains gave us some breathing space," Cameron told me during a lengthy interview in his window-lined office on a leafy street in Canberra, "and provided us with a real opportunity to plan for the next inevitable drought."

On other fronts, environmentally conscious farmers like Susan and Simon Tickner have learned how to adapt to the harsher weather. But the Tickners admit they couldn't have picked a worse time to give up their big jobs in Melbourne and move back to the farm Simon's family has owned in southeastern Australia for more than 70 years. "That was in 1998, and we essentially walked into 12 years of drought," Susan recalled. "The hardest thing was watching our first crop die when the drought hit."

But the couple miraculously posted a profit in subsequent years, even as farmers in dire financial straits around them were forced to leave fields fallow or kill emaciated livestock, by adopting a suite of strategies to squeeze out greater yield from an increasingly hostile, dry landscape—"more crop for the drop" is the catchphrase. "We're continually getting smarter about the things we can control," said Susan Tickner, "such as what mix of crops we put in, what nutrients we apply to maximize productivity, and how to cut costs and be more efficient."

Their shrewdest move was to adopt no-till farming, a practice in which seeds are planted without using a plow to turn the soil. This soil-saving farming method retains the carbon content of the soil, minimizing the amount of carbon dioxide spewed into the atmosphere. Plus, the healthy topsoil and stubble from the previous year's crop act like a canopy of mulch, with the decaying organic matter retaining moisture in the soil and providing nutrients and fertilizer to plants and preventing soil erosion. "Under the layer of dirt and mulch, there is this whole factory of microorganisms that make the crop grow and keep the soil healthy," Tickner explained. "If you burn or plow up your stubble, you're effectively breaking up that factory."

Having GPS navigation on their tractors enabled them to plant seeds within three centimeters of the target, so they were able to place them between the standing stubble. The knife point press wheel system they used, which left about a foot between the rows, kept soil disturbance to a minimum. They also used crop-management modeling software like the Yield Prophet, which was devised by Australian scientists to generate better yields given different scenarios of seasonal rainfall and temperature in response to climatic changes. Planting when the soil is dry, rather than waiting for the rains, saves fuel because it takes less energy to cultivate a dry field and it also gives them a jump on the growing season.

"There's no doubt in my mind the climate is changing," said Tickner, who belongs to a loosely knit cooperative of like-minded farmers called Climate Kelpie. "Since we've been on the farm, we're now planting at least four weeks earlier—and that is in just 13 years. People must be nimble—they'll have to change and adapt to various conditions to remain successful."

Technologies devised by Australian scientists can help, too. These include new farming systems to reduce water seepage, polymer ground covers to channel water better, and crops with genetic traits that are better suited to an environment that is hotter and drier and has elevated carbon dioxide levels. These research programs also serve as models for farmers in nearby developing countries, including the Philippines and Indonesia, which are anticipating widespread famine as the weather heats up and water becomes even scarcer. "Flexibility will be key, along with a willingness to explore the full menu of options," said Mark Howden, an agricultural science chief at CSIRO in Canberra and Australia's leading expert on how farmers can adapt to climate change. "In the future, they'll have to alter what they're doing on a continual basis" and be prepared to swiftly change factors such as the time of planting and the variety or species of the crop grown, and even to move to places where rainfall is more reliable.

It's not just agriculture that will suffer as the planet warms. The incidence of infectious diseases will rise, too, especially in tropical regions. Australian scientists are looking at ways to short-circuit the spread of disease; one that's in development is a novel method of combating dengue fever by biologically altering the mosquitoes to prevent them from transmitting the disease-causing microbes. If their strategy proves successful, it could create an entirely new way of stopping not only dengue, but also other bug-borne scourges, such as yellow fever, West Nile virus, Lyme disease, and even malaria.

"It could be a game changer," said medical entomologist Scott Ritchie. The jeans-clad, American-born scientist is part of the Australian research team that artfully engineered those insects. "It's almost like a dengue vaccine for the dengue mosquito," he told me one morning in April 2011. We were driving down a wide, palm-lined street in the Cairns suburb that in 2009 was the site of the largest dengue fever outbreak in Australia in 50 years. More than 1,000 people were stricken by the potentially fatal infection in an epidemic that swept through the beach towns of northern Australia with lightning speed.

"Right over there is where it started," Ritchie said, pointing to a green clapboard cottage perched on stilts to avoid flooding. The mosquitoes that transmit dengue congregate in the cottages' aboveground cellars, which provide damp breeding grounds. The summer of 2009 was especially hot and wet, and the strain of dengue that was involved incubated especially fast—perfect conditions for an epidemic. "The first patient contracted the virus in Indonesia and never sought medical care. By the time we detected it," he sighed, making a sweeping gesture with his arm, "it had spread throughout this whole area." And it was here, at Ground Zero, where the altered insects that could revolutionize how we fight bug-borne illnesses were first field-tested.

Stopping those diseases has never been more urgent. In North America, mosquitoes are pesky critters largely kept out of our lives by window screens and air-conditioning. But in hotter tropical climes, especially in the overcrowded megacities of Africa, Asia, and South America, the tiny insects are airborne angels of death, delivering lethal pathogens that sicken half a billion people and claim at least a million lives annually. Worse, mosquitoes are becoming resistant to the latest generation of pesticides, while decades of costly research to concoct vaccines for fatal ailments such as malaria have had limited success. The number of people vulnerable to mosquito-delivered afflictions is expected to soar into the billions as these disease-carrying strains extend their range into newly warm habitats—including in the United States.

At Ritchie's lab at James Cook University, on a densely foliaged campus near Cairns that is surrounded by rain forest–covered mountains, the team went through the tedious process of infecting mosquitoes with a bacterium called *Wolbachia* to prevent the bugs from passing on dengue. *Wolbachia* is a microbe present in 70 percent of insects. Scientists don't understand exactly why their Trojan horse strategy stops insects from transmitting the disease, they just know it works.

Their first step was coaxing female mosquitoes to lay eggs, which they only do after eating a blood meal (the researchers took turns getting bitten—which was harmless because the mosquitoes weren't carrying the dengue-causing virus). Only females drink blood, so only females transmit the diseases. Once sated, the mosquitoes were put in test tubes, where they laid their eggs within about an hour. The eggs, which look like tiny coffee grains, were placed on swatches of a red feltlike material. Then, scientists used ultrathin needles to inject the delicate mosquito eggs with the *Wolbachia* bacterium, which insinuates itself inside the bug. When the eggs mature into larvae, they're incubated in ceramic pots covered with mesh and stored on the gleaming stainless steel racks that line the lab. Once the larvae hatch, the cycle starts again. When the infected female mosquitoes reproduce, they pass the bacterium on to their offspring. And if an uninfected female pairs up with an infected male, said Ritchie, "he shoots blanks"—and the eggs fail to hatch—"giving the infected insects an extreme survival advantage."

Consequently, when altered mosquitoes are released in the wild, they should perpetuate themselves. In January 2011, Australian researchers tested that hypothesis by releasing about 300,000 infected bugs in two Cairns suburbs affected by the 2009 dengue outbreak. Within five months, the bacterium infiltrated the entire mosquito population and was soon affecting bugs in surrounding neighborhoods. Scott O'Neill, an entomologist at Monash University in Melbourne and head of the Australian team, told me the results exceeded all expectations.

Their next step will be to make sure this strategy actually stops the bugs from infecting humans. In lab cages, a slightly different strain of *Wolbachia* did prevent mosquitoes from transmitting dengue. In January 2012, researchers released 200,000 mosquitoes infected with this new strain in two other Cairns suburbs. The latest results show infection rates nearing 100 percent. But because Australia only has sporadic dengue outbreaks, the true trial will be in places like Vietnam and Indonesia, where dengue fever is endemic. The Australian team is awaiting government approvals to release lab-altered mosquitoes in those countries. "The beauty of this is that it's very green and should reduce our need for toxic insecticides," said O'Neill. "And it's cheap to implement and potentially sustainable in developing countries where these diseases are rife."

Parallel Universe

Traveling through the steamy northern tropics, in the flat grasslands of the bush in farm country, and in big cities like Brisbane and Canberra often feels to an American like being in a parallel universe. In the United States, many politicians scoff at climate change and attack the integrity of climate scientists. The hardy Australians, in stark contrast, are keenly aware that they face increasingly stormy weather, and they're buckling up for a bumpy ride. "You can't live in this area and not realize what is happening to the land," said Steve Turton, a physical geologist at James Cook University in Cairns who studies the effects of climate change on the country's rain forests and Great Barrier Reef. "My kids and their friends don't want to have kids because they can see what's coming."

Newscasters routinely link weather-related disasters to warming trends, a cabinet-level department is devoted to climate change, and the government has invested more than $500 million—a big chunk of change for a country of 22 million—in research on the impacts of climate change, including $126 million that's been earmarked specifically for projects to identify the best strategies for adaptation to a greater number of days of even hotter weather, storm surges, droughts, and rising seas. Australia's recently imposed carbon tax, which creates monetary incentives for curbing use of fossil fuels and investing more in renewable energy, is considered a model for other industrialized nations.

Australia is far from perfect: It is a coal "superpower," the world's largest exporter of this dirtiest of all fossil fuels. The island nation relies heavily on coal for electricity, and its per-person CO_2 consumption and agricultural emissions are among the highest in the developed world. "We are between a rock and a hard place," said Chris Cocklin, senior deputy vice-chancellor of research and innovation at James Cook University in Townsville. "We do have an economy that is deeply reliant on fossil fuel resources such as coal, but we are getting religion. With the right amount of will and leadership we can certainly avoid the worst of climate change."

To be sure, Australians took some convincing to get religion, especially since the deeply entrenched and powerful fossil fuel lobby was only too happy to use its tremendous economic influence to mount slick PR campaigns that sowed the seeds of doubt. But two pivotal events dispelled any lingering questions among most Australians about whether climate change was real and the ferocity of the threat they faced.

First there was the drought. While Australia has been ravaged by dry spells throughout its history, the recent decade-long drought, which Australians call "the Big Dry," was probably the worst since the Europeans arrived in 1788. The situation was so dire that in April 2007, then–prime minister John Howard suggested calling on a higher power. "We should all pray for rain," he said. Voters, who ranked sensible water management as one of their top priorities, were not amused and turned him out of office six months later.

Right after taking office, his successor, Kevin Rudd, under intense pressure to do *something*, appointed a commission to look at the economic fallout from a rapidly warming planet and what could be done to forestall the most severe consequences. The man he appointed to head the panel—Ross Garnaut, an economist at Australian National University, Australia's Harvard—had spent much of the previous three decades quietly advising a succession of Australia's leaders. Tall and slender with thinning gray hair and wire-frame glasses, the soft-spoken 60-something professor looked more like a kindly country doctor straight out of Norman Rockwell than a high-powered mandarin. But he possessed an astonishingly eclectic résumé.

Always staunchly probusiness and with deep ties to the mining, coal, and oil industries, he had advocated liberalizing banking rules, floating the currency, and removing trade barriers. He collected consulting fees from oil giants like Exxon and spent four years as ambassador to China, during which time he pried open trade between the two great Pacific Basin nations and generated investments in mining and smelting operations in Australia. Along the way, he'd served on the boards of nearly a dozen companies, organizations, and academic journals and chaired four of those businesses, including a gold mining outfit headquartered in neighboring Papua New Guinea.

Hardly the profile of a tree hugger, Garnaut himself admitted to being "agnostic" about global warming. Corey Watts of the Climate Institute described Garnaut as "a hard-nosed economic realist." Despairing activists and scientists can be forgiven for their cynical expectation that Garnaut's report would be a total whitewash. But he surprised everyone, including probably himself, and became, in his own words, "a late-life convert" to the green cause, and even seemed to revel in his startling transformation into an environmental rock star. His eponymous 2008 report echoed what climate scientists had been saying for years—that the growth in emissions will have severe and costly impacts on Australia's agriculture, infrastructure, biodiversity, ecosystems, and health, and that Australia is especially vulnerable because it is surrounded by developing countries like Indonesia and the Philippines that will be profoundly affected by climate change. But the fact that this came from a highly conservative economist silenced most adversaries, and the paper's conclusions were ominous and unequivocal: "There are times in the history of humanity when fateful decisions are made . . . the failure of our generation [to act on climate change mitigation] would lead to consequences that would haunt humanity until the end of time."

In a keynote address at the Greenhouse 2011 conference, a confab in Cairns that drew hundreds of scientists and activists from all over Australia, Garnaut painted an even grimmer picture. "There is a resistance in the emissions-intensive sector to change," he told his standing-room-only audience in a pointed jab at an industry that had once paid him handsomely for his advice. "But for people who are still skeptical, my question is 'When will you reach that point of desperation?' I'm not good at shouting and screaming," he continued, "and I understand the preference to wish it away. But if we do nothing," he added, and paused to look around the room for emphasis, "it will mean the end of civilization as we know it. We'll be one of the species facing extinction."

CHAPTER SEVEN

HOLDING BACK
THE WATERS

"All of this was underwater," recalled Billy Shanks, the New Orleans fire department captain. He gestured toward the two-lane road that runs along the placid waters of Lake Pontchartrain in New Orleans and to the two whitewashed yacht clubs and Landry's Seafood restaurant, a local landmark, perched along a finger of water that forms a tiny inlet a couple hundred yards away. It was a sultry June evening, and a cool breeze was blowing off the lake as the sun went down. We surveyed the serene scene while standing on the crest of a 20-foot-high earthen levee. The barrier didn't look like much—just a man-made pile of bulldozed dirt covered with a thin layer of grass—but the sloping hill created enough of an obstruction to protect the million-dollar brick and wood mini-mansions hugging the shoreline from storm surges off the lake. The vast, 640-square-mile body of water isn't technically a lake—it's an estuary linked to the Gulf of Mexico by a series of straits, and part of one of North America's largest wetlands. This neighborhood, known as Lakeview, had been uninhabitable swampland until the late 1940s, when a series of underground canals and pumps were installed to suck water out of the marshes and mounds of mud and concrete berms were built to hem in the lake.

But the levees notoriously failed on August 29, 2005, after Katrina swept through the city and blew a 200-foot hole in the 17th Street Canal—the largest and most important drainage canal in the city—which is a few blocks from where we now stood and not far from where Shanks lives with his family. "We got a call that there had been a breach," recalled Shanks. "Within about 20 minutes, the water had come up about 10 feet and flooded the entire area. We put on masks to come through on a boat—it was nasty-smelling stuff with dead, rotten animals and sewage, and it looked like black oil."

Today, there isn't a lingering trace of the damage; the entire community has been seamlessly reconstructed—including the faulty levees, which are now 12 feet higher than they were on that fateful day. The Shankses rebuilt their home, too, which had been swamped under eight feet of water, although it took nearly two years to cut through thickets of red tape to get financing. The butter-cream-colored two-story Craftsman is finished now, with gleaming hardwood floors and light-filled rooms that Flora Shanks lovingly decorated with garage sale and antique store finds, along with her own wood and glass artwork. The 17th Street Canal—a massive complex that consists of nearly a dozen 10-ton floodgates and 43 large drainage pumps—has been repaired as well, with a temporary pumping station and floodgates installed until the heavier artillery can be brought in and replaced with state-of-the-art equipment.

They're all part of a newly constructed levee system that became fully operational just in time for Hurricane Isaac, which blew in from the Gulf on August 29, 2012, drenching the coast with about 25 inches of rain and leaving more than a million people from Arkansas to Alabama without power. But New Orleans experienced relatively minor damage because the city is now like a walled fortress encircled by a sprawling, 350-mile-long system of linked levees, floodwalls, gates, canals, and pumping stations. The daunting $14.5 billion civil engineering project was completed in about six years, which is considered warp speed for an undertaking of this scale, and there is nothing even remotely similar to it in any other American city. Rebuilding the levees, however, is just one piece of an overall master plan to safeguard the city and the surrounding areas from floods and hurricanes. The other key element is a series of large-scale projects to restore Louisiana's deteriorating coastal wetlands, which serve as shock absorbers that disperse winds and prevent high storm surges like the one that filled the city when the levees broke.

We have about 100,000 miles of levees in the United States—enough to circle the globe four times—but most of these structures are old and dilapidated. In one of its annual infrastructure report cards, the American Society of Civil Engineers gave our levees a barely passing grade—D-minus—mainly because 85 percent of them were privately built more than half a century ago, and pegged costs for rehabbing these structures at upwards of $100 billion. Only about 35,000 miles of these levees are under some type of federal oversight, and the remainder—many of which are in disrepair or were poorly constructed—are managed by a patchwork of civil authorities that rarely even check whether they're trustworthy. Some are little more than mounds of soil. "Many of these levees started out as farmers piling up some dirt and running a grader down the road," said Gerald Galloway, a civil engineering professor at the University of Maryland. "The next year, they'd add a little more and pretty soon it's a levee protecting a housing development. Then some community group would take it over, but they don't have the money to adequately maintain it. The net result is that today hundreds of levees whose integrity are in question are in place in front of communities and properties."

Consequently, the grand plan to rebuild and safeguard New Orleans—which combines brick-and-mortar engineering marvels with refurbishing vital ecosystems to fortify the environment's natural defenses—could prove instructive for other parts of the country that are threatened by rising seas, drenching floods, and increasingly intense and frequent storms. Louisiana's ambitious restoration plan is one of many technological fixes being implemented or tested around the country that can help us adapt to climate change and stave off some of the worst effects of life on a hotter planet. These range from devising drought-resistant crops that can survive in harsher climes and smart-grid technologies that more efficiently manage energy consumption, to constructing comprehensive climate change infrastructure in vulnerable coastal cities and high-tech systems that transform sewage into sparkling, refreshing drinking water. What follows is a look at some of the more innovative projects that cities like New Orleans and Miami have launched to smooth the transition to a hotter future.

Rebuilding on Floodplains?

To be sure, it may seem like madness to continue living and building levees in these flood-prone regions, such as the Upper Mississippi River in Missouri and the California Delta. But as Willie Sutton once said when asked why he robbed banks, it's because that's where the money is. Settlements originally sprouted along rivers, lakes, and coastal regions because it just made the most economic sense: Waterways provide transit corridors for shipping goods, are the sources of drinking and irrigation water and fishing and other maritime industries, and enable pleasurable activities such as swimming and boating on hot summer days. While counties with levees account for only 37 percent of the nation's land area, they house more than 55 percent of the US population—more than 156 million people, according to a study commissioned by Levees.org, a New Orleans–based environmental advocacy group.

Louisiana is a good example of why communities flourish in floodplains. It's a relatively small state, but an economic powerhouse: The nation's largest supplier of shrimp and oysters, it ranks behind only Alaska in commercial fishing catches. It is the country's third-largest producer of petroleum, and contains about 10 percent of known US oil reserves—which means the nation enjoys lower energy prices because of the state's oil and gas production. More than 25 percent of the nation's waterborne exports, including more than 40 percent of all grains, pass through Louisiana's ports after they journey down the nation's largest river, the Mississippi, to where it empties into the Gulf of Mexico.

After Katrina, several reports, including an exhaustive $20 million investigation by the Department of Defense, blamed shoddy construction by the US Army Corps of Engineers for the massive breakdown of the city's levee system, declaring it "a system in name only" that was full of design flaws that made it vulnerable to collapse. "We do take accountability," said Lieutenant General Carl Strock, commander of the Corps, at a public meeting in downtown New Orleans less than a year after the storm ravaged the city. "It's been sobering for us to have to stand up and say we had a catastrophic failure of one of our projects."

A chastened Corps hunkered down and conducted a series of community forums to figure out where things had gone terribly awry, and how to fix it.

When the system was rebuilt, it was made heavily reliant on a $66 million network of floodgates and pumping stations at the ends of the outfall canals—the drainage conduits that siphon off water—that are capable of sucking up more than 50,000 cubic feet of water per second—that's enough power to fill up the Superdome in seven minutes. The system includes the West Closure Complex, a massive $1 billion pumping station—the world's largest—and surge barrier on the west bank of the Mississippi River, and a dramatically refurbished 17th Street Canal complex with concrete bunkers able to withstand winds of up to 250 mph and pumps that can sop up 4.1 million gallons of water per minute. The Corps also constructed the Seabrook Floodgate, a $165 million flood barrier spanning the Inner Harbor Navigation Canal shipping lane that links Lake Pontchartrain with the Mississippi and the Gulf Intracoastal Waterway. When storms threaten, two 50-foot-high gates can be lowered to seal off the canal from the kind of floodwaters that swamped the Lower Ninth Ward. About 12 miles east of New Orleans, a great wall about 26 feet high and nearly two miles long has been built across a corner of Lake Borgne. The massive barricade blocks the channel that funneled Katrina's worst storm surge from the Gulf into Lake Pontchartrain.

But some critics feel that the new levees are not strong enough to stave off disaster. The system is designed to protect against 100-year floods—which means that in any given year, there's a 1 percent chance this storm will hit. But Katrina was categorized as a once-every-400-years storm, and many climate change models predict storms of equal or even greater fierceness will become the norm in coming decades. In fact, major hurricanes could overwhelm the Mississippi River levees, according to a recent report by the Corps, which significantly increases the risk of floods. "The new levees are a little better, but they're like one of our pre–jet era planes—they just got new paint on the fuselage," said Robert Bea, a civil engineer at UC Berkeley and an expert in risk management. "It's a patchwork quilt with strong patches, but the weak ones were never touched. It will fail again, and fail severely."

WHAT WOULD THE DUTCH DO?

Out of sheer necessity, the Dutch have become the world leaders in flood-control management, and the Netherlands sets a standard of 10,000-year floods for its levee system, not just 100-year events. "The $14.2 billion fix is minuscule," said Bea, referring to the budget for the New Orleans levee system. "But to build Louisiana's levees to the Dutch standard would cost trillions, not billions. It would bankrupt the country." The Netherlands is small—about the size of Maryland—with about 16 million people, so it's one of the most densely populated places on Earth. More than two-thirds of the country is at or below sea level—in some places 20 feet below—and the Dutch have fought to hold back the water throughout their long history, dating back even as much as 800 years, when merchants and farmers created local water boards to "keep people's feet dry," as the locals explain it. Over the centuries, they built a vast network of levees, floodgates, ditches, canals, pumping stations, and storm-surge barriers that turned coastal swamps, shallow lakes, and low-lying bogs into productive land and transformed Amsterdam and Rotterdam, the latter of which has Europe's busiest port, into world-class cities.

But 60 years ago, on the night of January 31, 1953, a ferocious storm unleashed an 18-foot-high wall of water that bulldozed through the dikes holding back the North Sea and inundated the nation with its worst deluge in 400 years. The catastrophic flood killed more than 1,500 people, left 300,000 homeless, and caused more than $300 million in damage ($2.5 billion in today's dollars). While the drenched area made up only about 9 percent of Holland's territory, the psychological damage was intense and led the country to massively overhaul its North Sea defenses. What ensued was one of the most ambitious engineering projects of the last century, the Delta Works, which rivaled in scope the construction of our interstate highway system and the Hoover Dam. The Netherlands contains the deltas of some of Europe's largest rivers, including the Rhine, and the project involved building gargantuan dams and barriers in the southwestern corner of the country to permit shuttering the rivers and channels that funnel storm surges inland from the Atlantic Ocean and the North Sea. It took more than three decades to complete the massive project, and the Maeslant Barrier, the massive floodgates that safeguard Rotterdam, wasn't finished until 1996.

But global warming has ratcheted up the odds that even the Netherlands' reconstituted levee system will suffer a catastrophic breach. "The 1953 flood was our Katrina and jolted us out of our complacency," said Eric Boessenkool in a phone interview from his office in the Rijkswaterstaat, the Dutch water ministry, in The Hague. "But the barriers we built in the 1970s and 1980s are based on technology that we thought out 30 years ago. They won't protect us against climate change and sea-level rises of three feet or more by the end of this century. We don't have to panic because there is time, but we do have to get serious."

In 2006, the Dutch government launched an aggressive $100 billion construction campaign to climate proof the Netherlands for the next two centuries and prepare for waters that by 2200 may swell by as much as 13 feet. They've also shared their hard-won expertise on and technology for strengthening flood-protection systems, restoring the environment, and developing waterfronts in responsible ways with engineers, urban planners, and civic leaders in flood-prone regions of the United States, including Florida, Mississippi, New York, California, and Louisiana. Leaders in California and the Netherlands are investigating new approaches to strengthening existing levees and building systems to protect Central California, and to help make San Francisco and the adjoining Bay Area more resilient in the face of rising sea levels. In New York, in the wake of Sandy's devastation, Dutch officials are working with the Port Authority of New York and New Jersey to redesign the Brooklyn waterfront and discussing the feasibility of constructing massive floodgates near the mouth of New York Harbor and immense barriers in the Verrazano Narrows, the slim finger of water between Staten Island and Brooklyn, to fend off surges from superstorms.

But two decades ago, the Dutch learned some hard truths about the limits of constructing structures to harness nature. "In 1993 and in 1995, we had two major floods that forced hundreds of thousands of people to be evacuated, and we had no contingency plans," said Boessenkool. "We thought our levees were high and no one could hurt us anymore. But there was so much rain and the levees almost burst. It made us realize that to just keep raising the levees is not going to be a solution to our problems."

Climate change's melting of snowcaps in the Alps will bring Europe more winter rain, which will swell the Meuse and Rhine Rivers that flow into Holland. As a consequence, the Dutch instituted what they call the "Room for the River" project to allow flooding at dozens of locations along the

Meuse, Rhine, and Waal Rivers. They've moved dikes inland, and will peri-odically sacrifice some low-lying farming areas to floodwaters as a kind of safety valve that relieves pressure farther down the river, ultimately saving cities like Rotterdam from inundation. They're also looking at other ways of bolstering nature's protective mechanisms, such as linking together with earth and vegetation a chain of islands off the northern coast to form a buf-fer against the North Sea and converting some farms to marshy wetlands that are better able to absorb drenching rains.

Similar strategies are now being adopted in Louisiana because authorities there have come to many of the same conclusions: The walls protecting New Orleans can be built only so high. Dutch experts like Boessenkool, who has visited the Big Easy twice, are working with federal and state officials to restore marshlands to their natural state so they can serve as barriers during floods. Several projects, including the state's $50 billion Coastal Master Plan and the Army Corps of Engineers $3 billion project to shutter the Mis-sissippi River Gulf Outlet and restore the surrounding ecosystem, aim to bring the wetlands back to life.

Contrary to popular belief, experts now agree that the failure of the levees isn't what doomed New Orleans when Katrina made landfall, but rather the loss of this vital ecosystem, which could have greatly softened the blow of the storm surges. We now know that these marshes aren't just critical habitats for the flora and fauna that make the Louisiana bayous such exquisite and diverse tropical paradises. The intricate web of marshes, bayous, and barrier islands serves as a crucial land buffer that breaks storm surges and absorbs the energy of hurricanes. Storms lose their fury once they hit land because they aren't sucking up power from the overheated water, while the marshes are like sponges that soak up water, and the fric-tion from every single blade of swampland grass and every cypress tree immeasurably slows down the onrushing waters, thereby weakening the storm surge, and diffuses high winds. What scientists uncovered in look-ing at the 2004 South Asian tsunami was instructive—and in line with plain common sense. Coastal areas that were covered with mangrove for-ests sustained significantly less damage than the flat, sandy beaches in developed areas that offered no natural resistance to the series of deadly waves. Having similar barriers in Louisiana would have diminished the power and speed of Katrina and potentially averted the levee breaches that spawned the disaster.

But in the past 80 years, southern Louisiana has lost nearly 2,000 square miles of wetlands—an area about the size of Delaware—and the state has the dubious distinction of being one of the fastest-disappearing landmasses on Earth. Some of this is occurring because of natural geological cycles that are part of the settling of the Mississippi River Delta. But much of this is due to catastrophic soil erosion triggered by our misguided attempts to harness the Mississippi and by the hundreds of man-made navigation canals and oil and gas pipelines. Every year, about 30 square miles of marshes—a chunk bigger than Manhattan Island—vanish into the Gulf; that's about a football field of land every hour. By the time Katrina hit, the once thick blanket of thousands of acres of wetlands that had served as a lifesaving shock absorber to the south and east had been completely shredded, opening up an alley for water from the Gulf to head directly to New Orleans.

The disappearance of the swamps also creates a self-perpetuating cycle of destruction because the loss of natural protective barriers means the area is more susceptible to damage from storms: More salt water is able to intrude farther into the wetlands, more vegetation is demolished by rains and high winds, and more marshlands sink into the Gulf. Hurricane Rita struck the same area a few weeks after Katrina, and the two storms collectively wiped out more than 220 square miles of wetlands, as well as much of what was left of the chain of barrier islands that sheltered New Orleans. "What's frightening is there was more wetland loss due to those storms than we had expected over the next 50 years," said Gregory Miller, chief of the Plan Formulation Branch at the US Army Corps of Engineers in New Orleans. "Eighty to 90 percent of the wetlands loss that is going on in the nation is happening right on the Louisiana coast. It's an environmental problem of national significance."

How this happened is a veritable primer on how to unwittingly ravage a vital ecosystem. For starters, when the Mississippi River flooded during spring rains, the results of these deluges were devastating—so local, state, and federal agencies built dams and levees, which profoundly altered the river's plumbing system. The levees did their job and stopped the river from overflowing its banks, but straitjacketing the Mississippi had unintended consequences. It dramatically lowered the amount of sediment that would normally wash across the landscape during floods, naturally replenishing the soil. The upriver dams, in Minnesota and North and South Dakota, held back nearly a third of the sediment the river used to carry downstream, and the levee system formed a natural funnel that shunted much of the remaining marsh-building sediment into the Gulf. "In the 1700s, we started building levees on the banks all the way up the Mississippi," said Gerald Galloway. "We fenced in the river, and took the sediment that has been used in years past to maintain the wetlands. We did it for good and proper reasons—just like we denuded the Everglades—and at the time it seemed to make a lot of sense. But now we know better."

Over the past 60 years, engineers also constructed about 8,000 miles of canals and pipeline corridors for oil exploration and transporting drilling barges in the wetlands, along with roads and railways, accelerating the destruction of the ecosystem. These channels amplified soil erosion, interfered with normal drainage pathways after heavy rainfall, and created conduits that allowed salt water to seep in from the Gulf, which killed off native plants in the freshwater swamps. To make matters worse, in 1968, the Army Corps of Engineers completed dredging a 550-foot-wide shipping channel known as the Mississippi River Gulf Outlet (MRGO)—colloquially, "Mr. Go"—that was designed as a shortcut for freighters from the Gulf to the port of New Orleans. But the little used 76-mile transport corridor turned out to be a poorly conceived and costly multimillion-dollar ditch that displaced more dirt than the Panama Canal and allowed lethal doses of Gulf salt water to flood the marshes, killing off freshwater cypress and tupelo trees and destroying more than 30,000 acres of coast wetlands. ""It's very likely this project never would have been built if we understood the true environmental impact," said Aaron Viles, deputy director of the Gulf Restoration Network, a New Orleans environmental group. "The channel allowed an amazing intrusion of salt water that drastically changed the ecosystem and created a funnel effect that amplified and concentrated storm surges. Anywhere from 30 to 60 percent of our coasts is disappearing because of the oil and gas industry. Their canals were death by a thousand cuts. But MRGO was death by one huge cut."

Climate change and rising sea levels put these low-lying areas in even greater danger. Over the next 50 years, if nothing is done, the worst case scenario projects that water could swell by as much as three and a half feet, essentially drowning the marshlands and turning New Orleans into an island. "We don't want to see the open water of the Gulf of Mexico lapping on the edge of our town because then, any big weather event will put a lot of stress on the system," said Miller of the Corps.

After a series of probes into the culprits behind Katrina's devastation, Congress in 2006 decided MRGO should be shuttered and ordered the Corps to devise a plan to resurrect vital ecosystems. "The channel itself is a small part of a larger landscape these storm surges travel across," said Miller. MRGO has since been sealed off to shipping traffic and the Corps is hoping to begin work on a comprehensive plan that would nurse back to health about 60,000 acres of habitats that were decimated by the navigation channel. The stepwise project would start with the Central Wetlands on the eastern edge of New Orleans, about a 15-minute drive from the French Quarter, and gradually refurbish the fringe of barrier islands that form a natural barricade against the elements. Already, an alliance of environmental and civic groups have been working with water boards in the parishes around New Orleans to pump fresh and nutrient-laced sewage water into the Central Wetlands to hasten the growth of freshwater vegetation. Once these plants take root, the next step is to plant the mainstays of the ecosystem: cypress and tupelo trees.

Early in 2012, the Corps, working in tandem with cities and parishes in Louisiana and the state's Coastal Protection and Restoration Authority, released a comprehensive master plan that combines building higher levees with coastal restoration to protect New Orleans and other coastal communities from Category 5 hurricanes. Among the key provisions of the plan are projects that would fill in decimated barrier islands and natural ridges along the coast with sediment dredged mainly from the ocean floor, forming a necklace of land that could absorb some of the energy of storm surges. Gates and openings—called diversions—would be cut into levees along the Mississippi, which would allow revitalizing freshwater and sediment to wash over the wetlands and nurse these swamplands back to health. If these wetlands revival projects can get some traction, they will gradually slow the rate of loss, although it will take decades to turn the corner and begin to actually build land. By 2042, if the proposed plans are fully implemented,

Louisiana will begin gaining rather than losing land, and within 50 years, about two and a half square miles will be added annually.

"We face immense challenges with climate change and sea-level rise," Miller told me. "New Orleans and the surrounding coastal communities are built right on the edge for a number of reasons. We're at the mouth of the nation's largest river, the Mississippi; our landscape is at a much lower elevation than other coastal areas; and the rising of the sea is exacerbated because it's happening simultaneously with the sinking of the land, so the relative sea-level rise will be higher and more dramatic. We're going to feel the effects of this here before we do in a lot of other places, both in other communities on the Gulf, like Galveston, Mobile, and Tampa, and on the East Coast as well, in Miami, Savannah, and Charleston. But what we're doing is on the cutting edge, and what we learn from here can be applied to other places."

SURVIVING STORMY SEAS

The gleaming glass and steel high-rises perched along the sparkling white beaches that form Miami Beach's skyline are an integral part of the resort town's tropical allure. I used to come down here as a child in the 1950s to visit my grandparents, in the days before Cubans fleeing Castro turned it into a Latin American city. Even then, it seemed very exotic to my young eyes, an escape from New York's frigid and bleak winters—hot, sunny, sticky, and lush, with huge expanses of mangrove swamp flats filled with mosquitoes, alligators, and snakes. The streets were lined with palm trees and gleaming white buildings, and bulldozers seemed to be on every corner, adding dirt to what was left of the marshes to build upscale resorts and million-dollar mansions along the waterfront and Biscayne Bay. Today, however, the city sits on the front line of climate change, and the fabled global hot spot—long a playground for the ultrarich and famous, with its swanky hotels and condos and the trendy clubs, boutiques, and restaurants of South Beach—may soon be swallowed up by the ocean.

Most of Miami is just a few feet above sea level, and its location at the tip of a peninsula and its high population density make it especially vulnerable to the whims of unforgiving weather patterns. Even a three- or four-foot rise in sea level, which is entirely conceivable by the end of this century given current projections, could submerge Miami Beach and downtown Miami. Faced with a swelling ocean, heavier rainfall, and fiercer hurricanes that could cripple the economy and endanger residents, Miami is now more economically vulnerable to global warming than any other city in the world, according to research by the Organization for Economic Cooperation and Development, even more so than New York, New Orleans, or even Mumbai or Shanghai, with $3 trillion in assets at risk today and a projected $35 trillion by 2070. Vital infrastructure in the urban areas could be flooded or washed into the sea, including power plants, power generating facilities, schools, and hospitals. Each day, tons of international cargo from Europe, the Caribbean, and Latin America pass through Miami International Airport and the Port of Miami, both of which could be inundated. Miami's main source of drinking water, the Biscayne Aquifer, is also imperiled, as are its sandy beaches, wetlands, and parts of the nearby Everglades. In the not too distant future, residents of a hotter, wetter, and waterlogged Miami-Dade County might be forced to migrate to higher ground and settle along what will become a chain of islands but were once part of an ancient limestone ridge along the coast, abandoning their pricey beachfront dwellings to serve as reefs.

The urban landscape isn't the only thing that will be irrevocably altered by rising temperatures. All of South Florida is in grave danger—a four-foot sea-level rise would place 2.4 million people, 1.3 million homes, and 107 coastal cities at risk; soak nearly half the land; submerge most of the Florida Keys; and turn the Everglades into a saltwater bay. The rising sea level is already causing salty ocean water to intrude into those sizeable wetlands, killing off freshwater mangrove marshes, pushing salty bogs farther inland, damaging farmland, and tainting coastal freshwater wells, which threatens the drinking water of millions. Miami Beach's streets now flood often, not from drenching rains, but rather from underwater storm sewers overflowing during high tide; in the fall, when high tides are more frequent, roads may fill with water twice a day. During extreme high tides, some streets become impassable. "If climate change proceeds unmitigated," warned a recent report issued by the city, "living in Miami will become extremely difficult, if not impossible."

Civic leaders at the local and state levels, including Florida's previous governor, Charlie Crist, are keenly aware of the threat. Miami-Dade has been at the forefront of instituting numerous programs to prepare for an uncertain future. They include community revitalization projects to improve green spaces and mass transit, increase the availability of fresh and local produce through farmer's markets and partnerships with local growers, and to create pedestrian-friendly thoroughfares that encourage walking and cycling. The city also has introduced smart-growth initiatives to throttle greenhouse gas emissions, better methods of preventing beach erosion that have become widely copied models, and stricter building codes to make structures more energy efficient and able to withstand hurricane-force winds, storms, and floods. Through the Southeast Florida Regional Climate Change Compact, Miami-Dade is partnering with three other southern counties—Broward and Palm Beach, which have coastal communities on the Atlantic as far north as Palm Beach, and Monroe, which makes up the southwestern tip of the state and the Keys—to figure out how to short-circuit the worst effects of a warming planet.

Unlike New Orleans, Greater Miami doesn't have a buffer of wetlands to protect it from the rising seas, and, because of its geology, it can't simply build levees to hem in the ocean. The surface bedrock underneath Miami is limestone, which is porous. Below this layer of rock and soil lies the Biscayne Aquifer, which covers 4,000 miles and extends from the southern end of Palm Beach County to Florida Bay. But it is a shallow coastal aquifer, less than 20 feet below the surface, which means it is extremely vulnerable to contaminants and saltwater intrusions from the Atlantic and Biscayne Bay. "Building barriers is not a practical solution because of our hydraulics," said Nichole Hefty, climate change program manager for Miami-Dade County. "Essentially, you'd have to build barriers that would go hundreds of feet down below the groundwater. Otherwise, the water would flow under and around it."

In the 1940s and '50s, when much of South Florida was still mostly swampy wetlands and the area was undergoing the first of its many building booms, the Army Corps of Engineers constructed an extensive system of drainage canals and pump stations to control seawater intrusions and absorb floodwaters. But the increasingly outdated drainage system isn't up to the harsh future facing the region. After heavy storms, lingering floods of as much as three feet of standing water are now a fact of life in low-lying neighborhoods in South Beach, Miami Beach, and downtown Miami. Authorities have recently built 28 pump stations in the area, and the City of Miami is now considering a $206 million drainage system overhaul that includes construction of bigger and more powerful pump stations, raising seawalls, installing large storm water storage wells, and adding backflow preventers to drainpipes to stop salt water from seeping in from the Atlantic and Biscayne Bay.

Climate change also presents serious challenges for protecting the region's freshwater supply. Counties in South Florida have instituted a number of strategies to improve water-use efficiency and conservation measures that have reduced consumption by 20 percent over several years, and various projects are under way that will augment supplies with alternative water sources. Currently under construction is what will be the largest reclaimed water facility in Florida, which will purify wastewater and make it suitable for drinking. It is anticipated to produce 170 million gallons of water a day by 2027. "But all these projects are just the start of building a new infrastructure to cope with what's coming," said Hefty, as the region braces for increasingly stormy and unpredictable weather.

The Future of Water: Ground Zero

One sweltering day in September 2012, when temperatures in Los Angeles climbed into the triple digits and much of rest of the United States was soldiering through the worst drought since Eisenhower was president, I drove 40 miles south to Fountain Valley, a comfortable bedroom community in Orange County, to see firsthand the future of water. Tucked away on a side street in an industrial section of town near the freeway is the county's Groundwater Replenishment System (GRS), a $487 million high-tech solution to water scarcity where wastewater from toilets in Newport Beach that normally would be emptied into the Pacific after treatment is being recycled into water clean enough to drink. Every day, the sprawling 25-acre facility produces more than 70 million gallons of water, enough to quench the thirst of 600,000 Southern Californians, and construction is under way to increase the output by 30 million gallons a day by 2014.

Because water may soon be as precious as gold, Orange County's initiatives have become a widely lauded model of how to cope with anticipated shortages. More than 18,000 people from 20 different countries have taken tours of the water district's site, and its shelfful of awards, including the American Society of Civil Engineers' Outstanding Civil Engineering Achievement Award and the prestigious Stockholm Industry Water Award, which is considered the Nobel Prize of the water industry, attests to its world-class reputation. People are making the pilgrimage to this otherwise unremarkable Southern California suburb because the wastewater we normally flush down our toilets could provide us with a relatively cheap and plentiful supply in an era of increasing scarcity. If we recycled water that is normally dumped into estuaries or the ocean, according to a 2012 report by the National Academy of Sciences, we could boost US reserves by 27 percent, which amounts to about 12 billion gallons a day. The process is also less expensive than costly alternative technologies like desalination, which requires much more intensive and energy-consuming filtration systems to remove all the salt from seawater and is also limited to coastal states that can siphon off water from the ocean. And it costs less or about the same as importing water from distant sites, such as reservoirs or rivers, not just in actual dollars but also in the amount of energy needed to transport water—which is heavier than oil—in those aqueducts and pipelines.

Once the bastion of a Western-style political conservatism where Richard Nixon spent his final days—and home to Disneyland and John Wayne Airport—the affluent Southern California county is not typically viewed as part of the environmental vanguard. This region was once mostly farms and ranchland where cattle grazed lazily on thousands of acres of high desert and scrub brush. But the county underwent an explosive building boom in the middle of the last century, which saw the construction of thousands of homes in dozens of well-manicured planned communities that quickly mushroomed into a sprawling megalopolis of three million people—a number that is expected to double by 2050. When water managers peered into the future, it was obvious there wasn't going to be enough water to slake the county's growing thirst.

While Orange County is part of the same metropolitan water district as Los Angeles, officials felt their share of the water imported from the Colorado River and the Sacramento Delta was becoming too unpredictable and the dwindling supplies were making it increasingly expensive to boot (costs have doubled in the past 10 years alone). More than two decades ago—when the state was struggling through another of its periodic dry spells—water managers could clearly see the region was only going to get more parched and more congested. "We were coming out of a very severe drought and our board realized we needed to have a more reliable source of water," civil engineer Michael Markus, who is general manager of the Orange County Water District, told me during an interview in a conference room at the district's offices. "And with climate change, the trend lines were toward even drier weather and heavier storms, which makes rainfall more difficult to capture because the dams overflow. These were very serious problems and we could see that the only way to guarantee the water supply was through recycling wastewater. That way, we would have that control."

Civic leaders moved aggressively to make the county more self-sufficient through a combination of better managing local waterways and aquifers; constructing the Prado Wetlands, an artificial swamp that captures and naturally filters wastewater discharge from the Santa Ana River; and urban harvesting of rainfall and runoff. But the centerpiece of their effort was a massive technological fix: the Groundwater Replenishment System, the largest water purification project of its kind in the world. However, building the state-of-the-art treatment plant, which began in 2002 and was completed in 2008, was the relatively easy part. Their biggest obstacle was convincing residents to drink recycled sewage.

But Orange County's water honchos made a shrewd move at the outset that enabled their project to succeed when similar undertakings in cities like San Diego and Los Angeles were derailed by political opposition. They hired outside public relations consultants to figure out how to circumvent "the yuck factor," and developed a speakers bureau to spread the word. They held focus groups to identify consumers' concerns and made more than 2,000 presentations—to medical and health organizations, community and environmental groups, and local, state, and federal officials—in order to bring them on board. They even translated fact sheets into Spanish and

Vietnamese to develop trust among the people of these sizeable minority communities, most of whom emigrated from countries with poor water systems. Their successful outreach campaign has since become a template for success in other municipalities attempting similar projects, including a second try in San Diego and one in West Texas.

Orange County has also entered into partnerships with other governments doing the same thing in places like Australia and the Republic of Singapore to make the inevitable more palatable. Singapore's water woes mirror those of other cities. The capital city of the same name is now home to about five million people and has grown rapidly from a poor harbor town into an ultramodern business hub. But it is situated on an island that is far too small to provide enough water for its inhabitants. The city imports water from neighboring Malaysia, but the government doesn't want to rely on foreign sources and is smartly looking to a future when supplies may dry up. In addition to imposing strict water conservation rules to reduce consumption, capture storm runoff, and plug up leaks, Singapore has invested billions in high-tech water purification systems and embarked on a "NEWater" public relations campaign to make drinking reclaimed wastewater socially palatable.

"We had all these failures occurring around us and our board was very visionary in identifying that lack of public support would be a potential deal breaker," said Markus, a jovial man with a square jaw who supervised construction of the facility. "Our first letter of support was from the Centers for Disease Control, and that resonated very strongly with people we polled later on," he added. "We just pounded the community with presentations—whether it was audiences of 3 people or 300. We even talked to the Sons and Daughters of the American Revolution, anyone who would listen."

The recycling process begins at the adjoining Orange County Sanitation District, where solid wastes are removed and microorganisms are deployed to break down organic materials. The water is then piped to the GRS facility—and it still had a distinctive seaweedlike smell when I watched it gush into the vast tubs where it went through the first of a three-step purification process inspired by NASA technology for recycling waste on the International Space Station. Initially, the water is funneled through a series of straw-like microfiltration membranes that strain out organic carbons, bacteria, and tiny single-celled organisms like amoebas. Then it is shuttled through a reverse osmosis filter, in which water is pushed at high pressure through an ultrafine plastic membrane that removes even tinier organisms and compounds such as viruses, salts, pharmaceuticals, and pesticides. Finally, it is treated with hydrogen peroxide and exposed to high-intensity ultraviolet light to burn off any remaining impurities.

The system is carefully monitored 24-7 in a no-frills control room that looks more like the computer lab at a public high school than the nerve center of a state-of-the-art facility. "Automation is what makes this all possible because we can synchronize everything; we can step through cycles very quickly and optimize everything so it is energy efficient," William Dunivin told me as we stood in the middle of the spotless plant's floor, surrounded by the complicated filtration devices and powerful pumps that shunt millions of gallons of water through miles of massive pipes up to five feet in diameter every hour. Dunivin is director of water production and supervises the whole purification process. He's been with the water district since 1974 and has watched the technology evolve to the point where what they're doing is feasible. If anything goes wrong, the buck stops right here. But not much goes awry because of his obsessive attention to detail. "No one else is doing this at this scale," he said with no small measure of pride, "and we do a lot of in-house laboratory testing of next-generation technologies to constantly improve efficiencies. But the new phase," he added, motioning toward the construction site just beyond the plant's floor-to-ceiling glass windows—which is why we're all wearing hard hats and hazard vests—"will take this to the next level."

The water that emerges from all these fancy filtration systems is safe and eminently drinkable, clear, and refreshing, with the slightly mineralized taste of distilled water. But the public is still queasy about swallowing recycled sewage, no matter how many high-tech filters it passes through. So half of the reclaimed water is shunted 13 miles north to ponds in Anaheim where it slowly trickles down through sand and gravel to an underground aquifer that supplies drinking water for the county's residents. (The other half is funneled to a series of injection wells that line the seashore about five miles to the west to create an underwater dam that prevents briny seawater from seeping into freshwater reserves.) Even though studies show the recycled water is perfectly safe, the EPA has found people feel more comfortable drinking water that has passed through a natural purification process in a reservoir or an aquifer. "This is how we get over the 'toilet to tap' reluctance," Michael Markus told me. "We're looking at direct potable"—the nomenclature for the water that leaves the plant, skipping the last step—"although there's a lot of regulatory hurdles to overcome before we do that. That's what makes the most sense going forward. We're not letting our guard down, though, and we're still the best-kept secret in Orange County." But with water poised to become the oil of the 21st century, they won't be for long.

CHAPTER EIGHT

THE EMERALD CITY

After a week in New York City visiting an old friend, I went out for one last stroll before catching my flight from JFK to my life back in Los Angeles. At barely nine o'clock on a Sunday morning, the streets were virtually empty. But the sun had already burned off the early morning haze, the sky was dazzlingly clear, and the air was crisp with just a hint of an early spring chill. I made my way over to the High Line, the elevated urban park built on an abandoned rail line near my friend's apartment in Manhattan's picturesque West Village. Overlooking the Hudson River, the old freight line was erected in the early 1930s to prevent trains from colliding with street traffic, requiring that they be lifted more than 30 feet off the ground. The cars ferried meat, produce, and raw and manufactured goods from the factories and warehouses in what was once Manhattan's most industrial district.

But by 1981, the railroad had been abandoned and soon became a decayed eyesore of hulking iron that sliced through the west sides of Greenwich Village and Chelsea. The homeless camped under the trestle, and the rusting narrow strip was littered with debris and overgrown weeds, tall grasses, and wildflowers. Mayor Rudolph Giuliani slated it for demolition in the 1990s as the surrounding neighborhoods, including the Village's Meatpacking District, were rapidly gentrifying with art galleries, tony restaurants, chic designer shops, and million-dollar lofts with river views. But then a funny thing happened. A number of residents formed the advocacy group Friends of the High Line and orchestrated a savvy outreach campaign to keep it from being torn down—their glittery fund-raisers became hot tickets on the city's charity circuit—and then pushed for the transformation of the nearly seven acres of open space into a unique public place. It helped that Giuliani's mayoral successor, billionaire Michael Bloomberg, long a generous benefactor to the city's cultural institutions, thought it was a good idea, too.

At the time of my visit, the entrance at the southern terminus of the railroad, off Gansevoort Street, was surrounded by construction cranes, butcher shops, and dingy, aged warehouses. But up three flights of steel stairs, I found myself engulfed by an explosion of foliage—shrubs, flowers, trees, and tall grasses spilled over the protective metal railings on each side of the trestle, in a sudden and disorienting change of scenery. The park's walkway is composed of weathered wooden planks reminiscent of seaside boardwalks and flanked by steel-framed plant beds offering a wild prairie palette of muted hues—switchgrass, meadow sage, ironweed, wild quinine, and dozens of other species of native plants, even black-eyed Susans—that look natural, as if they were seeded by bird droppings or the wind. Yet all of them had been carefully selected because they're hardy survivors that can thrive in hostile urban conditions with shallow, infertile soil, little water because of the drying winds that blow off the river, and New York's harsh seasonal weather extremes. The designers went to great lengths to preserve the gritty, industrial architecture of the original space, with railroad tracks and ties embedded in portions of the pavement and landscape.

Farther down, the rail line threads its way through buildings that were once factories and then opens to a wide promenade where I bought a latte from a coffee cart vendor and sat on a sleek wooden lounge, drinking in the panoramic view of the Hudson. By midday, this urban oasis would be mobbed with thousands of locals and tourists, having become wildly popular beyond anyone's grandest expectations. But now it was deserted and eerily quiet except for the occasional squawking seagull perched on one of the rough-hewn wooden fences bordering the park's path. I savored this rare moment of tranquility when I could embrace the city with all its remarkable and beguiling contradictions without having to contend with the constant crush of humanity. To me, the reinvention of the High Line seemed an apt metaphor for New York, capturing the essence of the city: a matchless sanctuary of green against the backdrop of the Hudson, floating magically above the industrial hardscape and traffic in an urban center of eight million—which, despite its size, manages to be extraordinarily resilient and inventive while containing a sense of community, connectedness, and wonder.

When I had arrived at my friend's apartment earlier in the week, clumsily juggling my laptop and suitcase while fumbling with the keys at the building's front door, Rafael, the building's courtly and ultraefficient superintendent, was grinning broadly in greeting and patiently holding open the elevator door—even though it had been almost a year since my last visit. On our way up, we chatted amiably about the unseasonably warm April weather and about his son's impending graduation from a prestigious science and engineering college, which he had attended on a scholarship. Over previous visits, I had watched him mature from a gangly young boy into a studious and handsome young man. At that moment, I was struck by the fact that I knew more about Rafael's son than I did about the rambunctious kids who have lived across the street in our quiet little patch of paradise in Los Angeles for nearly a decade.

My friend's six-story apartment building was built in 1920, but the hallways are immaculate, and everything, even the ancient elevator, works with clocklike precision thanks to Rafael's meticulous care. Located on West Twelfth Street, it's right down the street from the White Horse Tavern, where Welsh poet Dylan Thomas took his last drink and across the street from Abingdon Square, a triangular half-acre wedge of greenery surrounded by a wrought-iron fence and covered with tall trees and a profusion of flowers. Here, people walk their dogs and flirt shamelessly with other pet owners, harried parents take their toddlers out for fresh air, and the elderly sun themselves on the benches that encircle the pocket park. On Saturday mornings, farmers travel from upstate New York for the greenmarket that's become a neighborhood institution and where regulars, including the celebrities in sweats and baseball caps who've driven area housing prices skyward, snap up heirloom vegetables, fresh-caught fish, artisanal cheeses, flowers, and organic baked goods.

Just a couple of blocks away is the townhouse on Hudson Street where Jane Jacobs wrote her highly influential 1961 masterwork, *The Death and*

Life of Great American Cities, in which she fiercely articulated her opposition to the urban renewal and roadway projects that were then sweeping the country. That orgy of postwar building focused on traffic flows and increasing population density, and it ravaged vibrant, well-functioning, but somewhat rundown neighborhoods and replaced them with stadiums, broad boulevards, and impersonal high-rises. Jacobs's beloved West Village had been deemed blighted and ripe for redevelopment with an expressway and tall apartment buildings, but she helped lead a successful fight to stop construction of an elevated highway in Lower Manhattan that would have destroyed her neighborhood. Even Eleanor Roosevelt became an ally.

At the time, she challenged many of the conventions of urban planning—and became the scourge of Robert Moses, New York City's enormously powerful master builder, in a David and Goliath struggle that became the stuff of legend. While Moses is now largely forgotten, Jacobs, who moved to Toronto in 1968 because she opposed the Vietnam War and worried about her two draft-age sons, has become a patron saint of the new millennium's urban renaissance. Instead of looking at the city as a squalid cesspool of inner-city decay, Jacobs saw just the opposite. She celebrated the urban sensibility, the rich street life, sheer density, and glorious chaos of the West Village, with its short blocks, cobblestone streets, low-slung buildings, historic brick townhouses, restaurants, taverns, bodegas, and family-owned businesses. The multiplicity of uses and exhilarating complexity encouraged walking and street traffic, and made the streets safer because they were never empty. But her greatest insight was that cities connected people to one another and created an intimate atmosphere that enhanced human interaction, which is what drives civilization's progress, and nurtured the casual connections that weave even a regular visitor like me into the rhythms and fabric of the city's life.

"The central idea of Jacobs's book is that density and diversity are the engines that make human communities work," noted writer David Owen in his insightful and provocative book *Green Metropolis: Why Living Smaller, Living Closer, and Driving Less Are the Keys to Sustainability*. "Society, she decided, has a critical mass. Spread people too thinly and sort them too finely, and they cease to interact; move them and their daily activities close together, and the benefits cascade: their neighborhoods grow safer, they become more attuned to one another's needs, they have more restaurants and movie theaters and museums to choose from, and their lives, generally, become more varied and engaging."

But for all her farsightedness, Jacobs could not have predicted that rowdy, boisterous, hypercaffeinated New York City in all its messy splendor—not the orderly, genteel, Birkenstock-wearing Portland or Seattle—would come to be considered one of the greenest cities in North America, ranking behind only Vancouver and San Francisco, because of the very qualities she extolled: its compactness and high density, which result in a tiny per capita carbon footprint. The Big Apple, with 27,000 people per square mile, is three times denser than any other city in the United States, which contributes to tremendous economies of scale in the use of resources. The comparatively little energy New Yorkers use is actually quite astonishing: Energy consumption on a per capita basis is about one third of the national average. While New York City encompasses nearly 3 percent of the US population, it accounts for only 1 percent of the nation's greenhouse gas emissions.

The city's gargantuan public transportation infrastructure enables more than a third of all New Yorkers to commute to work without using a private car—and in Manhattan, the figure rises to a staggering 76.7 percent. Because people don't drive, per person gasoline consumption is comparable to the national averages in the 1920s, when Calvin Coolidge was president and there were less than eight million cars on the road. Geography plays a key role, too: With 1.6 million residents jam-packed onto the tiny island of Manhattan, the only way to build is to go up, not out, which mitigates sprawl, translates into smaller home sizes that require less energy to heat and cool, and makes it easy to navigate on foot. This is also a borough with a huge park that encompasses about 850 acres of prime real estate—almost twice the size of Monaco—and even has its own zoo, which opened before Abraham Lincoln became president and underscores a historic civic commitment to valuing green spaces. Frederick Law Olmsted, considered the father of landscape architecture and the designer of Central Park, called public parks the "lungs of the city" that clear the air, absorb rainwater and pollution, and contribute to physical and mental well-being.

Sylvan paradises like Vermont, where you don't have to wait until farmers' market day to buy locally grown produce, may intuitively seem like places where sustainable living would be much easier than in urban areas. But the reality is quite different. Because the population is so spread out, Vermonters use nearly four times as much gasoline as New Yorkers, and six times as much as Manhattan residents: 546 gallons versus 90 gallons per person every year. Ironically, on just about every other barometer, Green Mountain State residents turn out to be the resource hogs: They have larger carbon footprints, guzzle more water, dump more garbage, and consume quadruple the amount of electricity as the average New Yorker. In other words, the seductive allure of rural life is simply wrongheaded at a time when the world's population is surging toward eight billion and roughly 80 percent of Americans live in cities.

It's easy to dismiss New Yorkers' remarkable energy efficiencies—after all, because they're all crammed together, they don't have much choice. But David Owen persuasively argues that's precisely the point: It is the unconscious efficiencies that are more desirable because they don't require enforcement or personal commitment. "The problem we face," he observed, "is how to make other settled places more like Manhattan, whose residents currently come closer than any other Americans to meeting environmental goals that all of us, eventually, will have to come to terms with."

We can't just sit back and rely on technology to help us adapt to this changing world and reduce the toll of disease from global warming. Many of the climate change mitigation strategies that will help us create a nonpolluting, clean-energy future also improve public health—they're what experts like to call "cobenefits." Promoting dense, public transit–oriented, pedestrian-friendly urban environments like New York not only cuts carbon emissions, but also generates multiple health benefits. This would decrease the incidences of obesity and chronic diseases such as heart disease and diabetes that are linked to excess weight, and curb air pollution, which is one of the culprits behind cardiovascular diseases, asthma, and other respiratory illnesses. Many eco-friendly cities are implementing sustainable development programs that are designed to cut traffic congestion, greenhouse gas emissions, and gasoline consumption in ways that include reducing the energy required to produce our food. They're also finding ways to encourage walking, hiking, and biking in the fresh air and sunshine to keep us fit and healthy and enhance our emotional well-being. "Health is not a minor fringe benefit," New York mayor Michael Bloomberg has noted. "The illnesses we can prevent through transportation and other policy changes that reduce combustion emissions—obesity-related, respiratory, and cardio-vascular problems—are among the most prevalent and expensive diseases in many developed countries. . . . A lower-carbon life leads to a healthier population."

Here again, New York is leading the way. Bloomberg has proved to be a responsible environmental steward, and his administration has taken bold steps to make the city even greener. On Earth Day 2007, he unveiled PlaNYC, a comprehensive, 127-point blueprint to create a sustainable city. It involves initiatives to revitalize the city's 580 miles of waterfront, decrease energy costs, cut greenhouse gas emissions by 30 percent by 2030, and upgrade the infrastructure to prevent shorelines from being swamped by flooding and storm surges. New York's temperature can be more than 7°F warmer than surrounding areas due to its intense urban heat island effect. By midcentury, experts predict the thermometer will spike by up to 5°F. More frequent and intense storms like Superstorm Sandy are on the way and will cause severe coastal flooding, while sea levels could swell by two feet or more, which would submerge most of the city's low-lying infrastructure, including roads and highways, subways and rail lines, and LaGuardia Airport.

In recognition of the fact that, as New York governor Andrew Cuomo observed after Sandy hit, "extreme weather is a reality," long-gestating plans to build three massive surge barriers—one on the East River, one near the Verrazano-Narrows Bridge, and another between Staten Island and New Jersey—along the lines of the Maeslant Barrier outside Rotterdam and London's Thames Barrier have been moved to the front burner, along with discussions about raising entrances to flood-prone subway stations and the locations of critical equipment like pumps, transformers, and wastewater pipes. The city has also launched several programs to turn down the climate's thermostat, such as the Million Trees Project (more than half a million have already been planted around the city) and painting flat roofs with a white or "cool" coating that reflects the sun's energy rather than absorbing it, as black roofs do. More than two million square feet of roofs have been covered so far, with the goal of coating them all by 2030. Heavy downpours can overwhelm New York's antiquated 19th-century sewer systems, the miles of pipes and tunnels that carry away wastewater and, when overtaxed, divert billions of gallons of raw sewage and contaminated urban runoff into the rivers, lakes, and ocean. The city has committed $1.5 billion to erecting green infrastructure—green roofs, street and sidewalk plantings, porous pavements, cisterns, rain barrels—that will control and absorb storm water and prevent polluted runoff from flowing into waterways by capturing it and filtering it naturally through the soil.

Protected bike lanes—separated in many places from street traffic by concrete curb barriers—have opened up all over the city, and New York now routinely appears on top ten lists of America's most bikeable cities. Helmeted riders can pedal relatively safely from Amsterdam Avenue in Washington Heights, near the northern tip of Manhattan, all the way down to the World Trade Center site and over the bridges into Queens and Brooklyn. New York's new Citi Bike program is expected to be the nation's largest bike share, with 600 stations and 10,000 bikes. Shared-use pedestrian and bike paths offering spectacular scenery run parallel to the Hudson River and the West Side Highway, and the city plans on having 1,800 miles of bike lanes by 2030. Perhaps not surprisingly, in a city where everyone walks, the percentage of people who are overweight or obese is 12 percent below the national average.

New York is also an acknowledged world leader in its efforts to create a more sustainable food system. A broad-based coalition of neighborhood activists, social justice and food advocacy groups, top chefs, writers, farmers, fishermen, ranchers, civic leaders, and state agencies have contributed to the development of a comprehensive suite of ground-to-garbage strategies to revamp the way food is grown, distributed, and disposed of. These methods could not only combat hunger and obesity by making more nourishing edibles more widely available, but also fight global warming by reducing energy use and better managing food waste. The improvements include ways to preserve and promote regional farming and food production; connect rural farmers with city markets; cultivate urban agriculture, including establishing rooftop farms and community gardens; create large-scale municipal composting facilities; and expand food cooperatives and outreach programs to residents in low-income areas, especially through the city's school food and after-school programs.

The hundreds of community gardens and dozens of farmers' markets that have sprung up in all five boroughs are part of the growing scaffold of a new food economy that will reduce the food footprint required to feed millions and ensure produce is fresher and more nutritious. That footprint includes not just the food we eat, but also the land, water, and energy required to grow, produce, and transport food to our plates. Food now often travels great distances, sometimes thousands of miles or more, through a complex network of warehouses and markets, highways and train stations, even container ships from Latin America and Asia. But New Yorkers are looking at ways to dramatically reduce these long food supply chains and eventually whittle down sources to those within a 500-mile radius of the metropolitan area, which will cut carbon emissions and improve quality. And what's happening in New York, one of the world's great megacities, can be a blueprint for other urban centers.

Bloomberg recently signed a package of new laws that promote, among other things, urban and rooftop farming initiatives that may soon become a model for sustainable local agriculture. In Brooklyn, to cite just one example, construction is under way on the world's largest rooftop farm in the 100,000 square feet of space on top of a former navy warehouse along the industrial waterfront. The massive eco-friendly greenhouse will grow produce hydroponically, a farming method that delivers nutrients to plants directly through water, without soil. The water will be recycled, which eliminates the fertilizer and pesticide runoff from conventional farming that is a leading cause of global water pollution. Plus, the consolidated space makes maintenance and harvesting easier and more efficient, which means that no gas-guzzling heavy equipment is needed. Once it is up and running, the farm should yield one million pounds of vegetables each year, enough to feed 5,000 New Yorkers.

New York State is home to more than 36,000 farms and seven million acres of farms. Consuming locally grown edibles is greener and healthier because produce loses about 40 percent of its nutritional value within three days of harvesting. That means that eating apples from farms in the nearby Hudson Valley is better than eating ones that spent several days en route from Washington State. But smaller outfits face a host of challenges to remain afloat; many have been forced into bankruptcy while real estate developers snap up the cheap land. In an area like the Catskills, the bucolic mountains 100 miles north of the city that are dotted with resorts, vacation homes, and bohemian artist colonies, there is no agricultural infrastructure—no creameries for dairy farmers, or meat processing for livestock, or quick-freeze facilities to preserve fruits and vegetables for shipping elsewhere. But civic and community leaders are working with Catskill farmers—"we're beyond the feel-good stage," said Mark Izeman, an environmental attorney who directs the New York urban program for the Natural Resources Defense Council—to hammer out concrete ways to revive the region's agricultural community.

On other fronts, plans are afoot to renovate the Hunts Point Produce Market, the world's largest produce wholesaler, which supplies fruits and vegetables from 49 states and 55 countries to feed the 23 million people who live in the metropolitan area. Sitting on 105 acres in the South Bronx and encompassing more than a million square feet of interior floor space, the mammoth facility was built in 1967 and is long overdue for a major overhaul. Increasing storage space, adding a wholesale farmers' market, and expanding the rail service deeper into Upstate New York to better connect farmers to the city and reduce the environmental fallout from the tractor-trailers that truck foods can help modernize the food-delivery hub and prepare it for a hotter 21st century.

"We're losing 70 acres of farmland every day to development, which adds up to a million acres paved over by concrete and asphalt in the last 60 years," said Izeman of the NRDC. "Creating a wholesale farmers' market for local farmers and better distribution networks for medium-sized and smaller farms sends a signal to local farmers they can make the investment to increase their production if they know there is demand. If we can resolve the distribution and infrastructure problems, we can make preserving farmland more economical."

Who would have thought it? That New York, known for its gritty street life and decaying 19th-century infrastructure, would come up with the right recipe for a healthier, brighter, more sustainable 21st century? Mix together sensible high-density living with a good public transit system, navigable streets, patches of green, and dynamic neighborhoods, then add in the foresight and political will to invest in infrastructure changes to prepare for the new realities of a hotter world and a carbon-free future and an active, engaged citizenry to push civic leaders to do the right thing—there's a special place in heaven for argumentative, loudmouthed New Yorkers—and voilà: the Emerald City.

THE DAMAGE DONE

New York developed as a city before the advent of the automobile, so it is compact and dense. To become more like New York, the rest of us are going to have to undo the half century's worth of damage to our health and the social fabric of our lives that resulted when we became a car-centric society and suburban sprawl became a way of life. Some forward-thinking cities have taken tentative steps in this direction—mainly the West Coast triumvirate of Seattle, Portland, and San Francisco—but most urban areas remain trapped in a car-reliant culture that isn't bad just for the environment, but also for our health.

In looking at how American society evolved in the past 60 years, it's almost as if some malevolent force had set out to create a dangerous, soul-crushing ecosystem that would make us miserable, lonely, and fat—and ravage the earth. We sit cocooned in our cars, detached from the reality of the world around us as we drive to work, run errands, and transport our kids to soccer practice, and we are usually too tired and too frazzled to get regular exercise. We live on isolated cul-de-sacs in cookie-cutter subdivisions and don't interact with the people who live next door, much less on the next block. This contributes to a steady erosion of social capital, those nourishing connections with our neighbors that create the web of shared community and nurture civic engagement and mutual trust. People like Jane Jacobs saw this coming, but they were largely ignored—until quite recently.

"We live in communities designed for automobiles and not for salving the human spirit, and we are only now beginning to understand the toll of these choices on our bodies," said Dr. Richard Jackson when I talked to him in his comfortably cluttered office at UCLA, where he is chair of the Department of Environmental Health Sciences. "The modern America of obesity, inactivity, depression, and loss of community has not 'happened' to us. We legislated, subsidized, and planned it this way. Thoughtful people now think this was a mistake, some sort of psychiatric break that we had in the 1950s, when we destroyed our cities with highways. But in truth, the connection between urban planning and health had been explored extensively a hundred years before, but we forgot because we'd get in our cars and drive away and not think about it."

Jackson, who coauthored *Designing Healthy Communities*, a companion piece to the PBS television series of the same name that he hosted, is one of a growing legion of public health experts who've become convinced that urban sprawl is at least partly responsible for many of the diseases that plague Americans, ranging from asthma, diabetes, obesity, heart disease, and hypertension to depression. They believe that the way we've arranged our so-called "built environment"—where we work, live, play, and shop—is profoundly toxic, and one of the chief culprits in disability and death in the 21st century.

The 60-something Jackson is a pediatrician by training and has the puckish looks, patience, and energy of someone who cares deeply about children and their future on this planet. Long a pioneer in the study of the complex interplay between our environment and our health, he had his personal "aha moment" on a steamy July morning in 1999, when he was director of the National Center for Environmental Health at the CDC in Atlanta. He was late for a meeting with the institute directors about a paper on the 10 leading diseases of the 21st century because he was stuck in traffic on Atlanta's notorious Buford Highway. The seven-lane thoroughfare cuts through a neighborhood of mainly poor immigrants that has no sidewalks and stretches of up to two miles between traffic lights and crosswalks. It was a beastly day, with temperatures hovering in the mid-90s. As he inched along, Jackson was struck by the sight of an elderly woman stumbling along the road's shoulder, bent over from osteoporosis, and weighed down by a plastic shopping bag in each hand.

Later, as he sat listening to his colleagues talk about the future of public health, he couldn't get the image of that struggling woman out of his mind. If she had collapsed at the side of the road, her death would have been attributed to heatstroke, he thought, and if she had been sideswiped by a truck, the cause of death would have been given as "motor vehicle trauma." But what would have really killed her was poor urban planning—too many cars, too much air pollution, a lack of sidewalks and public transportation, the black tar road that radiated heat, and the absence of trees and green spaces to cool down the air—and failed political leaders who neglected to rectify the situation. "Here I was focusing on remote disease risks," he recalled, "when the biggest risks that people faced were coming from the built environment."

A few years earlier, Jackson had moved his family from Berkeley, a woodsy, self-contained college town on the San Francisco Bay, to take the CDC job. But life in the Atlanta metro area, which has more than 16,000 miles of roads, covers 8,000 square miles, and has more than doubled in physical size since 1980 to accommodate its exploding population, proved a nightmare of urban sprawl. It took him an hour to drive the 10 miles to work, his wife drove an hour in the opposite direction, and he had to ferry his teenage son to the park on weekends to run track because there were no sidewalks in his neighborhood, the roadway was filled with potholes, and cars whizzed by at high speeds. "There was no way I was going to let him run on those awful, dangerous streets," said Jackson of the circumstances that catalyzed his career-altering epiphany.

"I'm a pediatrician and I'm a father," he continued. "I know the best kind of environments to rear children in—they need to explore, touch, grapple, and climb. But we've engineered physical activity out of our lives and out of our children's lives. We've created environments that don't work for children, and they don't work for families, and they don't work for everybody else."

UNINTENDED CONSEQUENCES

How we got here offers a textbook case of unintended consequences—in this instance, of well-intentioned societal changes that went seriously awry. People began the migration away from crowded, squalid industrialized cities at the turn of the last century, when trolley cars and ferries made commuting possible, moving out of the filthy, congested slums that were breeding grounds for infectious diseases and into environments with clean air and water. The suburban shift began in earnest in the 1920s, with more widespread ownership of automobiles and cheap and abundant fossil fuels.

But the trend greatly accelerated with the postwar prosperity of the 1950s, fueled by the construction of the interstate highway system and an extensive network of freeways and arterial roadways to link downtown business districts to the suburbs. (These were exactly the types of urban renewal projects that Jane Jacobs presciently found so heinous: They carved up established neighborhoods that once had had deep community roots, and as the middle class fled to the suburbs, only the poor remained in the hollowed-out urban cores.)

The highway system that changed America had its origins in World War II, and in particular, in the invasion of northern France by the Allied troops on D-Day in June 1944. Transporting troops, tanks, and equipment through Normandy's densely packed terrain and narrow roads proved to be a dangerous, arduous slog that made the invading army easy targets for the Axis forces, resulting in hundreds of unnecessary casualties and almost derailing the Allied crusade. But when the Allies finally broke through into Germany, they were able to access that nation's extensive autobahn network. General Dwight Eisenhower, the Supreme Allied Commander, was astonished by the greatly enhanced mobility afforded by the Nazis' extensive highway system, with its four-lane divided highways, on- and off-ramps, and no traffic lights, and the ability to swiftly move massive convoys of troops and equipment.

America lacked such a national highway defense system. When Eisenhower was elected president in 1952, he made the construction of one the signature issue of his administration because he believed we needed a network of highways that could "meet the demands of catastrophe or defense, should an atomic war come" to replace the patchwork of rutted country roads that spanned the nation's midsection. An engineering marvel considered by the American Society of Civil Engineers to be one of the Seven Wonders of the United States along with such other structures as the Hoover Dam, the Golden Gate Bridge, and the Panama Canal, the comprehensive system of roadways became one of the economic engines that drove our postwar prosperity and was responsible for an increase of about 25 percent in the nation's productivity. The vast scope of the system is mind-boggling: It covers 42,795 miles, with the longest route, Interstate 90, stretching 3,000 miles from Boston to Seattle, while I-95 extends nearly 2,000 miles from

Miami to Maine. It connects virtually every major metropolitan area and thousands of smaller communities and tourist destinations, and includes 55,500 bridges, 104 tunnels, 14,750 interchanges, and no traffic lights to accommodate the high speeds.

But in the name of progress, a way of life was lost and the adventure and romance of the road celebrated by footloose Beat Generation writers like Jack Kerouac, William Burroughs, and Allen Ginsberg were relegated to the dustbin of history. Once thriving and diverse businesses and communities along the fabled Route 66 died out, bypassed by the monotonous multilane interstate; they were replaced by motel chains, big-box stores, and fast-food franchises that made every place look like every other place. "It makes it possible to go from coast to coast without seeing anything or meeting anybody," Charles Kuralt, the late *CBS News* reporter, once tartly observed.

Huge subdivisions popped up seemingly overnight as the nation's population fanned out from the cities, transforming once pastoral small towns into bustling metropolises in a relentless quest to fulfill the American Dream and find housing they could afford. "Drive till you qualify" was the motto that eventually translated into insanely stressful and isolating commutes, some of three hours or more. Real estate developers dug up millions of acres of forests and farmland—which reduced local food supplies and strengthened industrial agriculture—and replaced them with chaotically designed, noisy suburban housing tracts that were far removed from the commercial zones where people worked and shopped, creating heavily car-dependent cultures. If residents wanted to go anywhere, even to pick up milk or a newspaper, they were forced to get behind the wheel. People who couldn't drive—the young, the elderly, the infirm, or the poor—became utterly reliant on others for rides to satisfy even the most basic needs because public transportation had not kept pace with development. They often found themselves homebound, socially isolated and depressed, another destructive but almost invisible side effect of our love affair with the automobile.

Today, significant numbers of people must still use patchy public transit systems and travel by foot in populated areas that are no longer built to accommodate pedestrians. The combination of wide suburban arterial roads, no sidewalks, few crosswalks, and high traffic speeds have made navigating some streets perilous. In one eye-opening study, researchers found that one-half of people ages 65 and over lived within a half mile of their food stores and pharmacies. But only 5 percent made the trips there on foot. A 2008 AARP survey of people over 50 revealed why: Almost 40 percent of those polled did not have adequate sidewalks in their neighborhood, nearly 47 percent could not cross their main roads safely because walk signals were too short and traffic was too fast, and 48 percent lacked a comfortable place to sit and wait for a bus. The unfortunate upshot of this is that Americans are far more sedentary and walk less than residents of any other industrialized nation, according to a 2010 University of Tennessee study, a fitness gap that reveals one cause of our burgeoning obesity epidemic. And in a self-perpetuating vicious cycle, because we're walking less, even fewer accommodations are made for pedestrians when streets are planned.

What happened to Raquel Nelson underscores just how dangerous traveling on foot can be. When the 30-year-old Atlanta college student arrived home with her three young children after a shopping trip to the local mall, it was dark because they had missed an earlier bus, which runs only once an hour. The bus deposited her across the street from her apartment complex in Marietta, Georgia, on the far side of a four-lane artery. The nearest crosswalk was three-tenths of a mile away. The weary family scooted across two lanes of busy traffic and stood on the median with other pedestrians waiting for a break in the traffic to cross to the other side. That's when the unthinkable happened—one of Nelson's daughters darted across safely, but her four-year-old son, who'd followed his sister, was killed by a hit-and-run driver.

Every month, about 400 pedestrians are killed in the United States, which is roughly equal to the number of passengers on a jumbo jet, according to a study by the Washington, DC–based nonprofit Transportation for America (TFA). An additional 688,000 pedestrians were injured in the decade from 2000 to 2009, a number equivalent to a pedestrian being struck by a car or truck every seven minutes. Two-thirds of all pedestrian deaths occur on busy thoroughfares that cut through residential communities, just like the one on which Nelson's son lost his life. One in four of all vehicle-versus-pedestrian crashes occurs within 100 feet of a transit stop. Those who survive after being struck may be severely incapacitated, even paralyzed. Often, they're knocked off their feet and strike their heads on the pavement or the car's hood, which can result in permanently disabling head injuries. "Americans get to pick their poison: less exercise and poor health, or walking on roads where more than 47,000 people have died in the last ten years," the TFA report concluded.

Perhaps not surprisingly, the worst pockets of these fatalities are in Sunbelt cities like Atlanta, Houston, Phoenix, and, in Florida, Miami, Jacksonville, Tampa, and Orlando, the latter of which tops the list, according to the TFA. These places underwent unbridled low-density growth at a time when the car culture was in its ascendency. "The science of engineering highways got applied to city streets," said Barbara McCann, executive director of the National Complete Streets Coalition. This is an alliance of advocacy groups that promotes revamping roads to make them safer for everyone, including cyclists, pedestrians, and especially the elderly. "Streets were built for speed to get cars moving as quickly and smoothly as possible," said McCann. "That included making sure cars had a clear sight line and that there was nothing for them to run into—which translated into eliminating trees, parked cars, and accommodations for pedestrians."

But cities can become more pedestrian friendly. The Complete Streets movement has developed a comprehensive blueprint of design changes that can make roadways safer and more appealing, such as adding dedicated bike lanes and sidewalks, building roundabouts and narrowing roads to slow traffic, raising crosswalks, and creating median islands to protect foot traffic. These ideas have been embraced by more than 300 communities and cities across the country, including San Francisco, Chicago, Detroit, and Charlotte, North Carolina. "Retrofitting takes a long time and it's incremental, but the safety benefits are enormous," said McCann. "Crash rates go down between 18 and 45 percent when people drive slower and pay more attention."

Reengineering cities so they can accommodate people on foot and on bicycles will be particularly important in the coming years as baby boomers get older—more than 62 million American will be age 65 or older by 2025—because it will allow them to age comfortably in place rather than being stranded in their homes by unfriendly streets as their faculties decline. We know what needs to be done—it's just a matter of finding the political will to get it accomplished. "When I started 30 years ago, there were no design guidelines for planners and bike plans didn't exist," said Peter Lagerwey, an urban planner and the project manager for the widely lauded Seattle Bicycle Master Plan developed by Toole Design Group, a firm specializing in pedestrian and bicycle transportation design. "But that's changed dramatically. There are now national guidelines and examples of what other communities have done—no one wants to be first or second and no one wants to be the last. There's also an incredible amount of research out there related to crash reduction, so we're getting a lot smarter about what works and what doesn't."

Strathcona and the Butterfly Effect

For a primer on how to do it right, look no further than Vancouver, North America's uncontested leader in smart growth, and consistently ranked among the world's most livable cities. This Canadian metropolis, which is home to more than 2.3 million residents, is a living laboratory evincing the active transformation of a modern city into a greener, healthier, and more public transit–oriented urban village. Its story is instructive—although at times Vancouver can seem like one of those insufferable overachieving nerds in high school who killed the curve in every subject, leaving all others to eat their dust. Among northwestern cities, Vancouver has the highest population density; the greatest rates of cycling, walking, and transit riding; and the fewest cars per capita, plus the highest life expectancy, lowest teen birth rate, and lowest poverty rate.

Vancouver also leads the world in the use of hydroelectric power, which makes up 90 percent of its energy supply, and regularly uses renewables like wind and solar. Its pedestrian-friendly, high-density neighborhoods (14,000 people per square mile) are clustered around high-rises, multifamily dwellings, restaurants, groceries, shops, and pocket parks, and have wide sidewalks that encourage walking and short blocks that slow traffic. The extensive mass transit system eases traffic congestion and includes high-capacity diesel buses and electric trolleys, a network of light-rail routes, and the elevated train service known as the SkyTrain. There are hundreds of miles of biking trails. Because of all of these factors, Vancouver has the lowest per capita carbon emissions of any major city in the Western Hemisphere—and that includes New York City.

"But we're like an aging supermodel," said Mike Harcourt, the former mayor who played a key role in the city's conversion from a Pacific Rim seaport and mining and timber-trading hub to a European-style green city. "Being livable isn't enough—now we're moving on to the next stage to become the world's most sustainable city by 2020." Vancouver's action plan echoes that of New York: It focuses on even greater use of renewable energy, further upgrades to the existing transportation system, and expansion of the city's already considerable green space, including the world-class Stanley Park, which is larger than Central Park. The blueprint also encourages urban agriculture, rethinks sewer and wastewater management, and reduces greenhouse gas emissions per capita to the lowest in the world by promoting even higher-density living with smaller multifamily homes, especially along transit corridors in Vancouver's downtown peninsula.

"About three-fourths of people who live in and around downtown don't take a car to work—we'd like to get that up to 95 percent by the time the transit system is complete," said Harcourt, who noted that the city's last mayoral election offered a choice between green and greener candidates. "Transforming the way we do cities is the key to mitigating, adapting, and hopefully moving away from the brink of some terrible consequences of climate change. Hopefully, Vancouver will be a mecca for urban sustainability."

But the Canadian city didn't become an eco-friendly flagship by accident. It kept its neighborhoods intact and avoided the nightmare of suburban sprawl because it never built an extensive system of freeways in the 1960s, when every other major urban area was tearing up its central core. But it wasn't because Vancouver's civic leaders somehow had greater insight into the ultimate consequences of so-called urban renewal than their counterparts elsewhere, because the fact is they didn't, and bulldozers had been revving their engines, ready to rip up significant chunks of the city.

A persuasive case can be made that it didn't happen, at least in part, because of a true butterfly effect: In the mid-1930s, Mary Lee Chan, a Chinese Canadian woman with a strong stubborn streak, refused to marry the man her father had picked out for her. Mary had been born in Vancouver in 1915, but her father decided it would be easier to raise and educate his children in China because of persistent racism in Canada. After the family settled in Guangdong Province near Hong Kong, Lee Wo Soon, as she was then named, fell in love with the man who was her teacher, Wah Goh Chan, and she too became a teacher—and refused the arranged marriage. "My grandfather believed girls should be educated because this was the way of the future, but he never expected this," recalled her daughter, Shirley Chan. "My mother was a strong personality and stood up to him even though it went against tradition."

After World War II, in the wake of the revelations about the Nazi atrocities, a chastened Canada repealed its Chinese exclusion laws. China was in turmoil, torn by a civil war that would see the Communists sweep into power in 1949. Lee Wo Soon returned to Vancouver, pregnant with Shirley, and worked three jobs, mostly in the garment industry, to save enough money to send for her husband and their eldest daughter. Wah Goh joined her two years later, barely escaping before the victorious Red Army took over. They became Mary Lee and Walter Chan, and the devoted soul mates put down roots in largely Chinese Strathcona, the city's oldest residential neighborhood, settled in the late 19th century when Vancouver was a Wild West boomtown and the entry point for successive waves of immigrants. They toiled at three jobs each, building a life for themselves and their children. Walter juggled factory shifts while working at the general store owned by Mary's father and brother, which was also a social nexus where the close-knit Chinese community congregated and could buy imported goods and foodstuffs from their native land.

But in 1959, a San Francisco–based consulting firm suggested building a network of freeways through the inner city, prompting Vancouver to announce plans to raze vast swaths of Strathcona to make way for new apartment buildings and a highway along the nearby waterfront. Newspapers heralded it as the much needed cleanup of a derelict slum that would revitalize the city's stagnant central business district. "The plan was to demolish the community in three phases—they would disperse the Chinese, who would be relocated to public housing projects, which would break up extended families and destroy natural communities, and then they would turn the land over to developers," recalled Shirley Chan. "When my parents first heard about the renewal program they thought the government would get rid of the blight. 'Oh good,' they thought, 'we'll just put more money into our house.' They even hired people to make needed repairs. But they didn't understand that their block was scheduled for demolition."

By 1965, more than 15 blocks had been leveled, clearing nearly 60 acres of valuable real estate and displacing more than 3,000 mostly ethnic Chinese, who were warehoused in nondescript concrete high-rises at the outer edges of the city. When the final phase was set in motion in 1968, with plans to flatten 600 more houses, the Chans faced eviction. "You can't stop it, the decisions have already been made," Shirley, who was then a 21-year-old college student, told her parents. Undaunted, Mary Lee went door to door with a plastic shopping bag draped over her arm, dragging Shirley with her to translate, asking neighbors to sign petitions, join the protests, and make donations to a legal defense fund. The scholarly Walter, weakened by the genetic liver disease that prevented him from accompanying Mary and eventually took his life, wrote a steady stream of articles for the Chinese-language newspaper that rallied the community.

The Chans turned out to be formidable foes.

More than 800 people showed up at city hall to protest. At a meeting organized in October 1968 by the social planning department of the city, Mary and Shirley Chan met Darlene Mazari, who had been hired by the city to help people relocate. But when Mazari heard how much hardship this was creating and how Chinese families were being robbed of everything they had worked for all their lives, she joined forces with the opposition. The Chans' three-story Edwardian home quickly became the focal point for the Strathcona Property Owners and Tenants Association (SPOTA), which was formed shortly before Christmas of 1968. The movement eventually involved hundreds of people—many of whom didn't speak English and had once been too timid to take on the still very British and lily-white city hall—and they organized fund-raising events and lobbied local politicians. Even Jane Jacobs, who had recently decamped to Toronto, was drawn into the struggle.

Mazari also recruited Mike Harcourt, who was then a 26-year-old storefront lawyer fresh out of school. Shirley and Darlene Mazari brought Harcourt to a local community meeting, conducted mostly in Cantonese, that lasted about three hours. "I'd been practicing law for about six months and I didn't understand a word, but afterwards, Shirley and Darlene told me, 'You've been hired to stop the freeway,'" he recalled. "'You'll just be taking on the city government; the provincial government; the auto, oil, gas, and transportation industries; and the real estate developers. But other than that, don't worry about it.'"

Harcourt became SPOTA's pro bono legal advisor. He provided legal counseling to hundreds of elderly Chinese who had never in their lives challenged authority and had lived in Strathcona for decades, but were so incensed that they planned to stand in front of the bulldozers and mobilize busloads of protesters in front of city hall. In the face of such intense opposition, the city capitulated and Harcourt helped to negotiate the final agreement that led to the rehabilitation rather than demolition of the neighborhood. The successful fight to save Strathcona swept progressives into positions of political power, including Harcourt, who was elected to the city council in 1973, then became mayor in 1980, and later was premier of British Columbia. Shirley Chan served as his mayoral chief of staff for more than five years—the first woman in Vancouver history to hold that position—before going on to a four-decade-long career in public service. Walter succumbed to his illness at age 67 in 1973, while Mary Lee lived almost three decades more, until 2002. She remained active in her community well into her 80s, but never remarried. The Chan House, at 658 Keefer Street in Strathcona, is now designated a historical place by the Canadian government.

The new political order spawned by their against-all-odds victory was the catalyst for completely revamped zoning regulations that allowed housing, restaurants, and shops in the city's downtown, not just office buildings, and the creation of a central area plan that transformed old industrial sites into thriving residential communities like False Creek North and Coal Harbour, which were once rail yards and shipping centers on the Vancouver peninsula. Today, Coal Harbour, a centrally located neighborhood within walking distance of downtown and Stanley Park, offers magnificent vistas and a waterfront park, while False Creek North is a high-density community of about 50,000 residents that was an Olympic Village for the 2010 Winter Games athletes and boasts a soaring steel-and-glass municipal science center and a popular boat marina.

"The Strathona Property Owners and Tenants Association is the single most significant reason why Vancouver is such a profoundly livable city today," recalled Harcourt, who recently coauthored *City Making in Paradise: Nine Decisions That Saved Vancouver.* "It was our turning point—otherwise

Vancouver would have been another failed city with an elevated freeway that would have wiped out two heritage areas and split and scarred three neighborhoods, and the central business district would be another sterile place that emptied out at five o'clock. We became ground zero of a whole different way of looking at what a city can and should be, and initiated the notion of citizen engagement and livable cities rather than the old and now discredited urban model that chased people out of perfectly good neighborhoods and created dead downtowns."

A MEDICAL MARSHALL PLAN

When I returned from my walk along New York's High Line that morning, the car I'd reserved to take me to the airport was sitting in front of my friend's building. I grabbed my bags and soon the driver was speeding through the Midtown Tunnel that runs under the East River and along an elevated expressway, high above the rail yards and warehouses, into an industrial section of Long Island City, Queens. I grew up about ten miles from there in Bayside, a leafy, solidly middle-class suburb of New England–style ranches with shutters on the windows, stately trees, and tidy lawns, on the northern tip of Long Island bordering Little Neck Bay, a finger of water off the Long Island Sound. It was an older community with a small-town vibe, farmland until the early part of the 20th century, when people in the entertainment business built mansions on the bay and it became a colony for movie and stage stars. Even silent screen heartthrob Rudolph Valentino once had a home there. The streets were narrow and safe, built before World War II, when cars were still a rarity. I walked everywhere as a kid—to school, to my friends' houses, to church on Sundays on the other side of Crocheron Park, to the stores on Bell Boulevard, and to the corner three blocks away where my friends and I would catch a bus into Flushing to see a movie. When I was in high school, we'd take the Number 7 subway line into Manhattan and go to the museums and matinees on Broadway and shop for bargains on 34th Street and at Klein's department store on Union Square.

After a brief stint in a converted brownstone on the Upper West Side in Manhattan after graduate school, I headed west in the 1970s to reinvent myself in a relentlessly sunny place that hinted of endless summer, happy to shed the snow and rain and my drippy old self. Los Angeles seemed like the future then, an arid high desert landscape coaxed into life with stolen water and palm trees imported from Mexico a century earlier to entice settlers. It had a distinctly space-age vibe, with sleek, modernistic midcentury wood-and-glass homes that had floor-to-ceiling windows and jetliner views of the breathtaking expanses of the world's most beautiful beaches. Everything seemed clean and shiny and new, with fast freeways, fast cars, and movie stars, and brimmed with infinite possibilities in a brawny, brassy collection of suburbs. But today, even Los Angeles is reinventing itself by developing higher-density living and walkable streets in cities like Venice, Santa Monica, and Manhattan Beach, and now it seems almost like Bayside, except with newer houses, less greenery, and better weather.

As my plane climbed into the sky on its western trajectory, I looked out the window and watched as the patchwork of communities surrounding the airport grew indistinguishable, fading into a grid of highways and wide boulevards. I thought about how I grew to young adulthood in a brief, shining moment of equilibrium in the heady postwar era of American dynamism. This was before the devastating consequences of our cavalier addiction to fossil fuels, our wasteful consumerism and squandering of natural resources, and the price we've paid in our health and emotional well-being for despoiling nature and the environment became apparent in all their stunning clarity. We will have to fight harder than we've ever fought to change the way we live and sacrifice some of our creature comforts to create a better, more livable and sustainable world. We must fight back the way Jane Jacobs and Mary Lee and Walter Chan fought, as if our lives and future generations depend on it—because they do.

Dr. Richard Jackson's parting words after our lengthy interview percolated in my head. "We're talking about reversing three generations' worth of building decisions that have created these heartless, thoughtless built environments," he told me as I gathered up my tape recorder and notebook and prepared to venture out into the gridlocked rush-hour traffic surrounding UCLA. "We need more-accessible healthy food, more parks, more farmers' markets, more bike routes in towns, and so on. We need to develop and have visions of things that are healthy, that are beautiful, that are friendly to the planet, and friendly to our health."

I thought back to a similar juncture in our history, when the fate of
civilization was in grave peril and so many Americans put aside their
differences to unite and vanquish a common enemy. It was during the
darkest days of World War II, when the Allies were mobilizing troops in
preparation for what became known as D-Day, the Allied invasion of
Nazi-occupied Europe that helped turn the tide of the war, and London
had been virtually leveled after years of relentless pounding by the Luft-
waffe. When thousands of American soldiers came pouring down the
gangplanks of the vast military cargo ships that docked in Southampton,
the strategic port city 60 miles south of London, the British were awe-
struck by their sheer size: one strapping six-footer after another, as if
John Wayne himself had stepped off the movie screen with all of his can-
do swagger.

Many of these men were like my father and my uncles: first-generation
Americans whose parents—my grandparents—had bravely immigrated to
the United States in search of a better life, much like the Dust Bowl settlers,
which is perhaps the reason I felt such a profound connection to them. They
had come from the shtetls in Germany and Russia, fleeing the Tsarist
pogroms, or from places like the poor hill country surrounding Lucca in
northern Italy, where my mother's parents had been born and raised before
they took their lives in their hands as teenagers and boarded a boat in Genoa
bound for New York City. They were among the most patriotic of Americans, and nearly burst with pride when their sons enlisted in the fight to
defend their adopted homeland.

The grateful, war-weary Brits, whose spirits were lifted by the arrival of the
Americans after years of deprivation and loss, treated the Yanks like libera-
tors—which is exactly what they were. Looking back, the Allied victory
seems inevitable, with the sweep of history surely on our side. But the fact is
the scales could have easily tipped the other way if not for the United States,
in tandem with the sacrifices made by the Soviets, who sustained more than
26 million casualties battling the Nazis and whose army inflicted mortal
wounds to Hitler's war machine. Here in the United States, the entire country was solidly behind the war effort. After the German blitzkrieg conquered
Europe, President Roosevelt, in one of his famous fireside chats in December
1940, before the United States entered the war, vowed that America would

become "the arsenal of democracy" and supply the Allies with the munitions they needed to vanquish the Axis powers.

And that's just what we did.

Within three months, the Big Three automakers halted auto production and quickly retooled, churning out tanks, bombers, aircraft engines, big guns, artillery shells, and armored trucks. When men went off to war after the Japanese attacked Pearl Harbor in 1941, women and racial minorities, who had been denied these jobs just a few months earlier, flooded the factory assembly lines, a shift that planted the seeds of two vast societal changes—the civil rights movement and women's liberation—that would change America forever. And on June 6, 1944, General Dwight D. Eisenhower led a fighting force of 160,000 Allied troops that landed on the heavily fortified beaches of Normandy, France, with more than 5,000 ships and 13,000 aircraft standing by in support off the wind-whipped coast. More than 9,000 soldiers were killed or wounded on that 50-mile stretch of beach that day, but 100,000 survived, and it marked the beginning of the final campaign to defeat Hitler.

After the war, Congress doled out more than $13 billion—when that was real money—to fund the Marshall Plan, an unprecedented economic and military rescue effort that airlifted thousands of tons of food to hundreds of thousands of starving refugees uprooted by the war and rebuilt a decimated Europe. This laid the foundation for NATO and prevented the western half of the continent from being overrun by the Soviets. This farsighted and humane investment was the brainchild of General George Marshall, America's foremost soldier during World War II, who built and directed the largest army in history before becoming president Harry Truman's secretary of state. For his efforts, Marshall was awarded the Nobel Peace Prize in 1953, the only soldier ever to be so honored.

We *must* become that America again.

It may seem like political dithering and right-wing skepticism has short-circuited meaningful discussion about what we must do to move to a clean-energy future and to make the necessary preparations to adapt to a changed climate. But no other country has the resources to lead the world in an enterprise on a scale as vast as what is required to truly save us from catastrophe. The sense of shared sacrifice, common purpose, and civic duty that pulled America through the Great Depression, that enabled us to trounce the most ruthless war machine ever created and then helped to lift a war-shattered Europe out of the ruins, is what could save us again. But it's going to require a vast paradigm shift in the way we live. The intricate fuel supply chain that transports oil from fields in the Middle East to refineries across the globe to the pumps in our neighborhoods has evolved over the past century. We don't have another 100 years to create an alternative fuel source on a scale needed for billions. "If we don't make the hard choices, nature will make them for us," noted Thomas Friedman in *Hot, Flat and Crowded.* "Right now, the acute awareness of the true scale and speed of this problem remains confined largely to the expert scientific community, but soon enough it will be blindingly obvious to everyone."

It's easy to despair because of the partisan sniping that has paralyzed Washington and the continued deep denial about the seriousness of the imminent threat of climate change. But a few years ago, I heard Robert F. Kennedy Jr. speak, and his words provided me with a shot of optimism that carried me through some of the darkest days of my research because his inspirational vision of a clean-energy future seems so easily within our grasp.

On a balmy summer evening in the ballroom of a posh oceanfront hotel in Santa Monica, an audience comprised mainly of members of the NRDC greeted him with a sustained standing ovation. RFK Jr., with his angular features and toothy grin, bears a striking resemblance to his fabled father, although he's taller and more solidly built. He is closing in on 60 now—more than a dozen years older than his father was when he was assassi-nated—and he wears the mantle of his family heritage and his iconic name lightly, often referencing his father in his speeches in ways that underscore how he is carrying on his legacy.

An environmental attorney with six children of his own, one of Kennedy's earliest childhood memories was of flying with his father over parts of West Virginia that had been ravaged by strip-mining in the Appalachian coal fields. Once a verdant, pristine wilderness, all that was left was mile upon mile of a desolate deforested moonscape where mining companies excavat-ing for coal had carved deep gouges in the mountainsides. The father described to his then-10-year-old son how some of the poorest people in the country were being made poorer by a coal industry that operated with little conscience or restraint. "They are not only ruining the land," the elder Kennedy told him, "but they are taking away these people's livelihoods."

The images were seared in the child's mind. So it's probably no surprise he's spent much of his life battling with predatory corporations to protect the powerless and to avert similar environmental Chernobyls. His years in the trenches have given him keen insights into the environmental challenges. "The biggest crisis we now face as a world civilization is global warming," he told the audience. Our consumption of oil and coal "is the principal drag on our economy. We are borrowing a billion dollars a day mainly from nations that don't share our values in order to meet our energy needs. This has beggared a nation that once controlled half the wealth on Earth. It's catastrophically expensive."

Our addiction to oil has not only bankrupted the country, he pointed out, but also makes us sick, killing at least 60,000 people annually with particles released from burning coal and inhaled, damaging young children who live downwind of factories and have higher rates of learning disabilities, and causing any number of other harmful effects. "And the hidden subsidies we give the oil and coal industries," he added, "form the barrier that prevents more economic and healthy forms of energy from entering the marketplace."

Yet far from being daunted by the formidable challenges we face, Kennedy was energized. The tantalizing picture he painted was of highways filled with lightweight electric cars freed from hauling around 500-pound internal combustion engines. He envisioned a dazzlingly bright carbon-free future powered by low-cost alternative energy sources—solar, wind, geothermal—that will liberate us from our dependence on oil, unfetter an economy staggering under the weight of an exorbitantly expensive and inefficient fuel source, and rescue us, in his words, "from genuflecting" to the anti-American despots who bankroll terrorism. "Every nation that decarbonizes its society experiences instantaneous prosperity," he said. "We need to create a national marketplace that attaches to the grid and allows people to sell power back to the grid. We need to create a marketplace that rewards good behavior, which is efficient, and punishes bad behavior. Right now, we have a destructively perverse marketplace that rewards the filthiest, dirtiest, most destructively poisonous fuels from hell rather than cheap, clean, green, wholesome safe fuels from heaven. We need to change that around and create a market that allows every American to hook in and sell energy and the lowest cost provider wins. Every American becomes an energy entrepreneur and every home becomes a power plant. We need to power this country on human energy and imagination and what Franklin Roosevelt called American industrial genius rather than Saudi oil."

I thought about the dozens of committed people I'd encountered in my research, the dedicated scientists and public health doctors from all over the world, the hardy Aussies on climate change's front lines, the heroic physicians, nurses, and firefighters in New Orleans, the green activists in New York and Vancouver, and the environmental justice advocates here in California, all of them doing their utmost to make the world a better place. While 2012 was the warmest year on record—and the continental United States was gripped by unprecedented numbers of droughts, floods, and superstorms—what we are seeing is, quite literally, the tip of the iceberg. To meet these staggering challenges, we must harness their best ideas and use them to create a medical Marshall Plan. We must raise the scaffolding of a national public health infrastructure that is resilient and able to blunt the disastrous effects on our health and well-being as climate change intensifies. We have excellent models of what we need to do in cities: Las Vegas; Vancouver; New York; New Orleans, with its network of medical clinics and data storage banks that can be accessed even in a power outage; Philadelphia and Chicago, whose heat wave programs are now widely copied; and Orange County, California's innovative water reclamation technology.

We are on the threshold of transformative changes in which natural calamities will convulse the globe, possibly leading to unlivable cities, widespread famine, civil unrest and wars over dwindling resources, the extinction of at least half the species on Earth, the death of the oceans from growing acidity, and hundreds of millions of desperate people uprooted by drought, floods, fires, and other weather-induced catastrophes.

It doesn't have to be this way. If we make a concerted effort to curb carbon emissions, we can make the optimistic visions of people like Robert Kennedy and Richard Jackson a reality.

But we can't waste another minute.

NOTES

INTRODUCTION: WHEN THE WEATHER GOES HAYWIRE

Abbey, K. Black Sunday. *Amarillo Globe-News*, April 16, 2005. http://amarillo.com/stories/041605/new_1725542.shtml.

Amland, B. H. 42 million displaced by natural disasters in 2010. *Salon*, June 6, 2011. www.salon.com/2011/06/06/natural_disasters.

Begley, S. Are you ready for more? *The Daily Beast*, May 29, 2011. www.thedailybeast.com/newsweek/2011/05/29/are-you-ready-for-more.html.

Busch, M. Links between climate change and increased social unrest unacknowledged. Truthout, October 14, 2011. www.truth-out.org/climate-change-often-unacknowledged-contributor-increased-violence-and-authoritarianism/1318276701.

Climate Central. Top 10 states ravaged by extreme weather in 2011. OnEarth.org, December 20, 2011. www.onearth.org/article/top-10-states-ravaged-by-extreme-weather-in-2011.

Climate Institute. *A climate of suffering: The real costs of living with inaction on climate change*. Melbourne: Climate Institute, 2011. www.climateinstitute.org.au/verve/_resources/tci_aclimateofsuffering_august2011_web.pdf.

Collins, M., et al. The impact of global warming on the tropical Pacific Ocean and El Niño. *Nature Geoscience* 3:391–397, 2010, doi: 10.1038/ngeo868.

Cooney, C. Preparing a people: Climate change and public health. *Environmental Health Perspectives* 119:166–171, 2011, doi: 10.1289/ehp.119-a166.

Costello, A. Running a temperature: Climate change will soon be the world's greatest health crisis. *Foreign Policy*, June 24, 2009. www.foreignpolicy.com/articles/2009/06/24/running_a_temperature_0.

Derbyshire, D. Climate change is the biggest health threat of 21st century, claims report into global warming. *Mail Online*, May 14, 2009. www.dailymail.co.uk/sciencetech/article-1181355/Climate-change-biggest-health-threat-21st-century-claims-report-global-warming.html.

Earth Gauge. Drought and climate change. Fact Sheet, December 2007, www.earthgauge.net/wp-content/fact_sheets/CF_Drought.pdf.

Egan, T. *The worst hard time*. New York: First Mariner, 2006.

Erlich, B. Interview by J. Todd, July 17, 1986, Shattuck, OK [DVD]. Oral History Program Oklahoma Historical Society, 1986.

Friedman, T. L. *Hot, flat, and crowded: Why we need a green revolution and how it can renew America*. New York: Farrar, Straus and Giroux, 2006.

Friedmann, J. Why it's hard to talk about energy. *The Atlantic*, March 29, 2011. www.theatlantic.com/technology/archive/2011/03/why-its-hard-to-talk-about-energy/73147.

Garza, J. Interview by J. Todd, May 15, 1985, Boise City, OK [DVD]. Oral History Program, Oklahoma Historical Society, 1985.

Gore, A. Climate of denial: Can science and the truth withstand the merchants of poison? *Rolling Stone*, June 22, 2011. www.rollingstone.com/politics/news/climate-of-denial-20110622.

———. The scorched earth: Climate change could bring severe drought to much of the world within decades. *Rolling Stone*, June 24, 2011. www.rollingstone.com/politics/news/the-scorched-earth-20110624.

Hesterman, D. Stanford climate scientists forecast permanently hotter summers beginning in 20 years. *Stanford Report*, June 6, 2011. http://news.stanford.edu/news/2011/june/permanent-hotter-summers-060611.html.

Honigsbaum, M. Anthony Costello: Making climate change part of global health. *The Lancet* 373(9676):1669, 2009, doi: 10.1016/S0140-6736(09)60929-6.

Ignore climate change and 100m people will die by 2030, shocking new report claims. *Mail Online*, September 26, 2012. www.dailymail.co.uk/sciencetech/article-2208953/Shock-report-claims-100m-people-die-economic-growth-drop-3-2-2030-climate-change-ignored.html.

Jackson, B. Interview by J. Todd, September 12, 1984, Woodward, OK [DVD]. Oral History Program, Oklahoma Historical Society, 1984.

K101 Radio. Black Sunday [CD]. Oral History Program, Oklahoma Historical Society, 1985.

Knowlton, K., et al. Six climate change–related events in the United States accounted for about $14 billion in lost lives and health costs. *Health Affairs* 30(11), 2167–2176, 2011, doi:10.1377/hlthaff.2011.0229.

Lister, S. Professor Anthony Costello: Climate change biggest threat to humans. *Times Online*, May 14, 2009. www.timesonline.co.uk/tol/news/uk/health/article6283681.ece.

McKibben, B. Global warming's terrifying new math. *Rolling Stone*, July 19, 2012. www.rollingstone.com/politics/news/global-warmings-terrifying-new-math-20120719.

McManus, G. On "Black Sunday." *Tulsa World*, April 18, 2010. www.tulsaworld.com/site/printerfriendlystory.aspx?articleid=20100418_65_G3_Thedeb869826&PrintComments=1.

Paddock, C. Climate change biggest threat to health, major study. *Medical News Today*, May 14, 2009. www.medicalnewstoday.com/articles/150107.php.

Parfit, M. The Dust Bowl: Half a century ago, parts of the Great Plains blew away, and the question now is: Could it be happening all over again? *Smithsonian*, June 1, 1989. www.accessmylibrary.com/coms2/summary_0286-5454076_ITM.

Portier, C. J., et al. *A human health perspective on climate change*. Research Triangle Park, NC: Environmental Health Perspectives/National Institute of Environmental Health Sciences, 2010. www.ibcperu.org/doc/isis/12254.pdf.

Schubert, S. D., et al. *Long-term drought in the United States Great Plains*. Climate of the 20th Century Workshop, Calverton, Maryland, January 22–25, 2002. www.iges.org/c20c/workshops/200201/ext_abs/schubert_ext_abstract.pdf.

———. On the cause of the 1930s Dust Bowl. *Science* 303(5665):1855–1859, 2004. http://climate.envsci.rutgers.edu/pdf/Schubert2004Science1855.pdf.

Stallings, F. L. Black Sunday: The great dust storm of April 14, 1935. Austin, TX: Eakin Press, 2001.

Williams, T. Rare infection strikes victims of a tornado in Missouri. *New York Times*, June 10, 2011. www.nytimes.com/2011/06/11/us/11fungus.html.

Yosemite hantavirus warning expands to 39 countries. LA Now [blog], LATimes.com, September 5, 2012. http://latimesblogs.latimes.com/lanow/2012/09/yosemite-hantavirus-warning-expands-to-39-other-countries.html.

Zongker, B. Heat gripping half of US expected to last for days. *Times-Picayune,* June 9, 2011. www.nola.com/weather/index.ssf/2011/06/heat_gripping_half_of_us_expec.html.

CHAPTER 1: FEVER PITCH

In addition to the sources cited in the chapter, I'm especially indebted to Mary Hayden, a scientist with the University Corporation for Atmospheric Research in Boulder, Colorado, who did much of the arduous gumshoe detective work gathering information about the 2005 dengue outbreak in Brownsville, Texas. Kerry Clark, an expert in vector-borne diseases at the University of North Florida, and Dr. James Diaz of the Louisiana State University patiently explained to me how air travel has accelerated the spread of disease. Michael McGeehin and Lyle Petersen of the CDC illuminated efforts by the agency to identify and combat the health effects of climate change and to beef up disease surveillance networks because of alterations in ecosystems.

The following written materials were also used:

Acuna-Soto, R. et al. Large epidemics of hemorrhagic fevers in Mexico 1545–1815. *American Journal of Tropical Medicine and Hygiene* 62(6):733–739, 2000.

———. Megadrought and megadeath in 16th century Mexico. *Emerging Infectious Diseases* 8(4):360–362, 2002.

Alto, B. W., and Juliano, S. A. Temperature effects on the dynamics of *Aedes albopictus* (Diptera: Culicidae) populations in the laboratory. *Journal of Medical Entomology* 38(4):548–556, 2001.

Big, old mice spread deadly hantavirus. *Science Daily*, January 9, 2009. www.sciencedaily.com/releases/2009/01/090106230720.htm.

Centers for Disease Control and Prevention Dengue Branch. Update: Dengue in tropical and subtropical regions. CDC.gov, November 22, 2012. wwwnc.cdc.gov/travel/notices/outbreak-notice/dengue-tropical-sub-tropical.htm.

Climate change may alter malaria patterns. *ScienceDaily*, February 14, 2009. www.sciencedaily.com/releases/2009/02/090214162631.htm.

Dengue hemorrhagic fever—US–Mexico border, 2005. *Morbidity and Mortality Weekly Report* 56(31):785–789, 2007. www.cdc.gov/mmwr/preview/mmwrhtml/mm5631a1.htm.

Dumler, J. S., et al. Human granulocytic anaplasmosis and *Anaplasma phagocytophilum*. *Emerging Infectious Diseases* 11(12), 2005. wwwnc.cdc.gov/eid/article/11/12/05-0898_article.htm.

Grady, D. Death at the Corners. *Discover* 14(12), December 1993. http://discovermagazine.com/1993/dec/deathatthecorner320.

Gubler, D. J. The continuing spread of West Nile virus in the Western Hemisphere. *Clinical Infectious Diseases* 45(8):1039–1046, 2007.

———. The president's address: Prevention and control of tropical diseases in the 21st century: Back to the field. *American Journal of Tropical Medicine and Hygiene* 65(1)5–11, 2000.

———. Resurgent vector-borne diseases as a global health problem. *Emerging Infectious Diseases* 4(3):442–450, 1998. www.ncbi.nlm.nih.gov/pmc/articles/PMC2640300.

Hall, S. S. On the trail of the West Nile virus. *Smithsonian*, July 2003. www.smithsonianmag.com/science-nature/westnile.html.

Heinrich, J. *West Nile virus outbreak: Lessons for public health preparedness.* US General Accounting Office Report GAO/HEHS-00-180, September 2000. www.gao.gov/assets/230/229648.pdf.

Kilpatrick, A. M., et al. West Nile virus epidemics in North America are driven by shifts in mosquito feeding behavior. *Public Library of Science Biology* 4(4): e82. www.plosbiology.org/article/info%3Adoi%2F10.1371%2Fjournal.pbio.0040082.

Knowlton, K., et al. *Fever pitch: Mosquito-borne dengue fever threat spreading in the Americas.* Natural Resources Defense Council Issue Paper, July 2009. www.nrdc.org/health/dengue/files/dengue.pdf.

Larsen, K. The new diseases on our doorstep. *OnEarth*, Fall 2009. www.onearth.org/article/the-new-diseases-on-our-doorstep.

Linden, C. USAMRIID supports West Nile virus investigations. DCMilitary.com, October 5, 2000.

Nash, D., et al. The outbreak of West Nile virus infection in the New York City area in 1999. *New England Journal of Medicine* 344:1807–1814, 2001. www.nejm.org/doi/full/10.1056/NEJM200106143442401.

National Science Foundation. Ecology of infectious diseases: Medical mystery solved. n.d. www.nsf.gov/news/special_reports/ecoinf/solved.jsp.

Natural Resources Defense Council. New report: Dengue fever a looming threat in the United States [press release]. July 8, 2009. www.nrdc.org/media/2009/090708.asp.

Ogden, N. H., et al. The emergence of Lyme disease in Canada. *Canadian Medical Association Journal* 180(12):1221–1224, 2009. www.cmaj.ca/content/180/12/1221.full.

Petersen, L. R., et al. West Nile virus encephalitis. *New England Journal of Medicine* 347:1225–1226, 2002. www.nejm.org/doi/full/10.1056/NEJMo020128.

Ramos, M. M., et al. Epidemic dengue and dengue hemorrhagic fever at the Texas-Mexico border: Results of a household-based seroepidemiologic survey, December 2005. *American Journal of Tropical Medicine and Hygiene*, 78(3):364–369, 2008. www.ajtmh.org/content/78/3/364.long.

Ramshaw, E. Major health problems linked to poverty. *New York Times*, July 9, 2011. www.nytimes.com/2011/07/10/us/10tthealth.html.

Rawlings, J. A., et al. Dengue surveillance in Texas, 1995. *American Journal of Tropical Medicine and Hygiene* 59(1):95–99, 1998. www.ajtmh.org/content/59/1/95.long.

Richard, M. G. Climate change too abstract for you? Dengue fever could spread to 28 US states. Treehugger.com, July 8, 2009. www.treehugger.com/natural-sciences/climate-change-too-abstract-for-you-dengue-fever-could-spread-to-28-us-states.html.

Silverman, A., et al. Toronto Emergency Medical Services and SARS. *Emerging Infectious Diseases* 10(9), 2004. wwwnc.cdc.gov/eid/article/10/9/04-0170_article.htm. [letter]

Soverow, J. E., et al. (2009). Infectious disease in a warming world: How weather influenced West Nile virus in the United States (2001–2005). *Environmental Health Perspectives* 117(7):1049–1052, 2009. www.ncbi.nlm.nih.gov/pmc/articles/PMC2717128.

Steinhauer, J. African virus may be culprit in mosquito-borne illnesses. *New York Times*, September 25, 1999. www.nytimes.com/1999/09/25/nyregion/african-virus-may-be-culprit-in-mosquito-borne-illnesses.html.

Struck, D. Climate change drives disease to new territory. *Washington Post*, May 5, 2006. www.washingtonpost.com/wp-dyn/content/article/2006/05/04/AR2006050401931.html.

Stutz, B. Megadeath in Mexico. *Discover*, February 2006. http://discovermagazine.com/2006/feb/megadeath-in-mexico.

Timeline: SARS outbreak. CNN.com, April 24, 2003. http://articles.cnn.com/2003-04-24/health/timeline.sars_1_sars-outbreak-hong-kong-s-hospital-authority-mysterious-respiratory-disease?_s=PM:HEALTH.

US Global Health Policy Program. *The global malaria epidemic.* Fact sheet, March, 2013. www.kff.org/globalhealth/7882.cfm.

World Health Organization. Infectious diseases kill over 17 million people a year: WHO warns of global crisis [press release]. n.d. www.who.int/whr/1996/media_centre/press_release/en/index.html.

CHAPTER 2: BREATHLESS

Dr. David Pepper, a tireless fighter for environmental justice and one of the architects of Medical Advocates for Healthy Air in Fresno, spent many hours educating me about the long history of efforts to clean the air in the Central Valley. Fellow journalists Kari Lydersen and Rebecca Plevin generously shared their contacts and insights into the damaging effects of pollution at California's ports and in the Central Valley, while Elizabeth Jonasson of the Coalition for Clean Air in Fresno brought me up-to-date on where things stood now.

I also benefited from discussions with Bart Ostro of the California Environmental Protection Agency about his work on climate change and how rising temperatures affect mortality and hospitalizations. Conversations with pediatrician David Peden, who directs the UNC Center for Environmental Medicine, Asthma and Lung Biology at the University of North Carolina in Chapel Hill, illuminated the effects of air pollution on children's health.

These written materials also provided excellent background:

Anderson, B. Fleeing Valley's unhealthy air. *Fresno Bee*, December 12, 2007. www.fresnobee.com/2007/12/12/263223/fleeing-valleys-unhealthy-air.html.

———. Fresno is state's asthma capital. *Fresno Bee*, December 12, 2007. www.fresnobee.com/2007/12/12/263218/fresno-is-states-asthma-capital.html.

———. Researchers looking for answers. *Fresno Bee*, December 12, 2007. www.fresnobee.com/2007/12/12/263230/researchers-looking-for-answers.html.

Arenofsky, J. Valley fever blowing on a hotter wind. Scientific American.com, April 15, 2009. www.scientificamerican.com/article.cfm?id=valley-fever-hotter-wind.

Arizona washes away dust from storm. *USA Today*, July 7, 2011. www.usatoday.com/weather/storms/2011-07-07-phoenix-dust-storm_n.htm.

Balakrishnan, K., et al. (2004). Executive summary. Indoor air pollution associated with household fuel use in India: An exposure assessment and modeling exercise in rural districts of Andhra Pradesh, India. World Bank, June 2004. pp. 10–15. http://ehs.sph.berkeley.edu/krsmith/?p=228.

Bauerlein, M. When health pros fight back. *Sierra*, July/August 2006. www.sierraclub.org/sierra/200607/health.asp.

Begany, T. (2001). Study: Fewer cars equal fewer asthma exacerbations. *Respiratory Reviews* 6(5), May 2001. www.respiratoryreviews.com/may01/rr_may01_fewer.html.

Bernard, S., et al. The potential impacts of climate variability and change on air pollution-related health effects in the United States. *Environmental Health Perspectives* 109(Suppl 2):199–209, 2001. www.ncbi.nlm.nih.gov/pmc/articles/PMC1240667/pdf/ehp109s-000199.pdf.

Bilger, B. Hearth surgery. *New Yorker*, December 21, 2009. www.newyorker.com/reporting/2009/12/21/091221fa_fact_bilger.

Bunting, M. How Hillary Clinton's clean stoves will help African women. Poverty Matters [blog], Guardian.co.uk, September 21, 2010. www.guardian.co.uk/commentisfree/cifamerica/2010/sep/21/hillary-clinton-clean-stove-initiative-africa.

California Environmental Protection Agency Air Resources Board. California Air Resources Board approves advanced clean car rules [press release]. January 27, 2012. www.arb.ca.gov/newsrel/newsrelease.php?id=282.

———. *Facts about particulate matter mortality: New data revealing greater dangers from PM2.5*. May 30, 2008. www.ncuaqmd.org/files/Particulate Matter Mortality Fact Sheet.pdf.

Christopher, T. Can weeds help solve the climate crisis? *New York Times Magazine*, June 29, 2008. www.nytimes.com/2008/06/29/magazine/29weeds-t.html.

Coalition for Clean Air. Valley air quality groups urge air board not to delay clean air, propose solution to smog pollution crisis [press release]. April 25, 2007. http://ccair.org/news-desk/view-cca-news/press-releases/press-release-archive/266—valley-air-quality-groups-urge-air-board-not-to-delay-clean-air-propose-solution-to-smog-pollution-crisis.

Comrie, A. C. Climate factors influencing coccidioidomycosis seasonality and outbreaks. *Environmental Health Perspectives* 113(6):688–692, 2005. www.ncbi.nlm.nih.gov/pmc/articles/PMC1257592.

Cone, M. A toxic tour: Neighborhoods struggle with health threats from traffic pollution. *Environmental Health News*, October 7, 2011. www.environmentalhealthnews.org/ehs/news/2011/1008a-toxic-tour-of-la.

Cone, T. Valley fever on the rise in Central Valley. *San Francisco Chronicle*, February 13, 2009. www.sfgate.com/cgi-bin/article.cgi?f=/c/a/2009/02/12/BAC115T91A.DTL.

Dickson, E. C. "Valley fever" of the San Joaquin Valley and fungus coccidioides. *California and Western Medicine* 47(3):151–155, 1937. www.ncbi.nlm.nih.gov/pmc/articles/PMC1753515.

Dixon, D. Can climate change clean up indoor air? Scientific American.com, July 12, 2010. www.scientificamerican.com/article.cfm?id=can-climate-change-clean-up-indoor-air.

Donovan, G. H., et al. The relationship between trees and human health: Evidence from the spread of the emerald ash borer. *American Journal of Preventive Medicine* 44(2):139–145, 2013.

Earth Justice. Settlement reached on air pollution lawsuit in San Joaquin valley [press release]. EarthJustice.org, October 9, 2001. http://earthjustice.org/news/press/2001/settlement-reached-on-air-pollution-lawsuit-in-san-joaquin-valley.

Epstein, P. R. Fossil fuels, allergies, and a host of other ills. *Journal of Allergy and Clinical Immunology* 122(3):471–472, 2008. www.jacionline.org/article/S0091-6749(08)01359-6/fulltext.

Fann, N., et al. Estimating the national public health burden associated with exposure to ambient PM2.5 and ozone. *Risk Analysis* 32(1):81-95, 2012.

Fighting for air. *The Fresno Bee*, December 12, 2007. www.fresnobee.com/2007/12/12/263215/fighting-for-air.html.

Friedman, M. S., et al. Impact of changes in transportation and commuting behaviors during the 1996 Summer Olympic Games in Atlanta on air quality and childhood asthma. *JAMA: The Journal of the American Medical Association* 285(7):897–905, 2001. http://jama.ama-assn.org/content/285/7/897.full.pdf.

Gordon, S. Climate change could sting allergy, asthma sufferers. *USA Today*, May 11, 2009. www.usatoday.com/news/health/2009-05-11-climate-allergies_N.htm.

Graham, S. Global study links climate to rates of childhood asthma. ScientificAmerican.com, June 21, 2004. www.scientificamerican.com/article.cfm?id=global-study-links-climat.

Hennessy-Fiske, M. Dust storm shrouds Texas City. *Los Angeles Times*, October 18, 2011. http://articles.latimes.com/2011/oct/18/nation/la-na-dust-storm-20111019.

Jacobson, M. Z. Enhancement of local air pollution by urban CO_2 domes. *Environmental Science and Technology* 44(7):2497–2502, 2010. www.stanford.edu/group/efmh/jacobson/Articles/V/es903018m.pdf.

Jarrett, M. Kirk Smith brings global attention to the effects of solid fuel use on health and climate change. *Bridges*, Summer/Fall 2011. http://coeh.berkeley.edu/bridges/summer2011/stove_initiative.html.

Kirkland, T. N., and Fierer, J. Coccidioidomycosis: A reemerging infectious disease. *Emerging Infectious Diseases* 3(2):192–199, 1996. wwwnc.cdc.gov/eid/article/2/3/pdfs/96-0305.pdf.

Martinez, A. Port of Los Angeles proposes developing the cleanest truck program. Adrian Martinez [blog], Switchboard.nrdc.org, April 20, 2012. http://switchboard.nrdc.org/blogs/amartinez/port_of_los_angeles_proposes_d.html.

McKinley, J. Infection hits a California prison hard. *New York Times*, December 30, 2007. www.nytimes.com/2007/12/30/us/30inmates.html.

Michelozzi, P., et al. High temperature and hospitalizations for cardiovascular and respiratory causes in 12 European cities. *American Journal of Respiratory and Critical Care Medicine* 179:383-389, 2008. http://ajrccm.atsjournals.org/content/179/5/383.full.pdf.

Miller, K. A., et al. Long-term exposure to air pollution and incidence of cardiovascular events in women. *New England Journal of Medicine* 356(5):447–458, 2007. www.nejm.org/doi/full/10.1056/NEJMoa054409.

Mustafic, H., et al. Main air pollutants and myocardial infarction: A systematic review and meta-analysis. *JAMA: The Journal of the American Medical Association* 307(7):713–721, 2012.

O'Connor, A. Air pollution linked to heart and brain risks. Well [blog], NYTimes.com, February 15, 2012. http://well.blogs.nytimes.com/2012/02/15/air-pollution-tied-to-heart-and-brain-risks.html.

Sanders, R. Auto exhaust linked to thickening of arteries, possible increased risk of heart attack. University of California at Berkeley News Center, February 8, 2010. http://newscenter.berkeley.edu/2010/02/08/atherosclerosis_particulates.

Sanders, R., and Anderberg, L. Cutting greenhouse pollutants could directly save millions of lives worldwide [press release]. University of California at Berkeley, November 25, 2009. www.berkeley.edu/news/media/releases/2009/11/25_lancet_papers.shtml.

Schwartz, J. Air pollution and children's health. *Pediatrics* 113(Suppl 3):1037–1043, 2004. http://pediatrics.aappublications.org/content/113/Supplement_3/1037.long.

Stenson, J. Allergies trigger agony across the US. NBCNews.com, April 24, 2009. www.msnbc.msn.com/id/30267953.

Sunyer, J., et al. Urban air pollution and emergency admissions for asthma in four European cities: The APHEA Project. *Thorax* 52:760–765, 1997. http://thorax.bmj.com/content/52/9/760.full.pdf+html.

Tolme, P. Get ready to itch and sneeze. *Newsweek*, August 11, 2008.

Trasande, L., and Thurston, G. D. The role of air pollution in asthma and other pediatric morbidities. *Journal of Allergy and Clinical Immunology* 115(4):689–699, 2005, doi: 10.1016/j.jaci.2005.01.056.

University of California at San Francisco. UCSF Fresno starts infectious disease fellowship program [press release]. July 25, 2008. www.universityofcalifornia.edu/news/article/18289.

Vidal, J. Dust storms spread deadly diseases worldwide. *The Observer*, September 26, 2009. www.guardian.co.uk/world/2009/sep/27/dust-storms-diseases-sydney.

Vugia, D. J., et al. (2009). Increase in coccidioidomycosis—California, 2000–2007. *Morbidity and Mortality Weekly Report* 58(5):105–109, 2009. www.cdc.gov/mmwr/preview/mmwrhtml/mm5805a1.htm.

Wayne, P., et al. Production of allergenic pollen by ragweed (*Ambrosia artemisiifolia* L.) is increased in CO_2-enriched atmospheres. *Annals of Allergy, Asthma and Immunology* 88:279–282, 2002. http://chge.med.harvard.edu/sites/default/files/resources/pollen_wayne_epsteinpdf.pdf.

Wellenius, G. A., et al. (2012). Ambient air pollution and the risk of acute ischemic stroke. *Archives of Internal Medicine* 172(3):229–234. www.hsph.harvard.edu/clarc/sac2012/endnote290-wellenius.pdf.

Weuve, J., et al. Exposure to particulate air pollution and cognitive decline in older women. *Archives of Internal Medicine* 172(3):219–227, 2012. http://archinte.jamanetwork.com/article.aspx?articleid=1108716.

Wilkinson, P., et al. Public health benefits of strategies to reduce greenhouse-gas emissions: Household energy. *The Lancet* 374(9705):1917–1929, 2009. www.thelancet.com/journals/lancet/article/PIIS0140-6736(09)61713-X/fulltext.

Yang, S. Long-term ozone exposure linked to higher risk of death, finds nationwide study [press release]. University of California at Berkeley, March 11, 2009. http://berkeley.edu/news/media/releases/2009/03/11_ozone.shtml.

———. Wood smoke from cooking fires linked to pneumonia, cognitive impacts. University of California at Berkeley News Center, November 10, 2011. http://newscenter.berkeley.edu/2011/11/10/cookstove-smoke-pneumonia-iq.

Yeung, B. Southern Californians at risk of death from air pollution, EPA says. CaliforniaWatch.org, February 13, 2012. http://californiawatch.org/dailyreport/southern-californians-risk-death-air-pollution-epa-says-14843.

Ziska, L. H., and George, K. Rising carbon dioxide and invasive, noxious plants: Potential threats and consequences. *World Resource Review* 16(4):427–447, 2004. http://afrsweb.usda.gov/SP2UserFiles/ad_hoc/12755100FullTextPublicationspdf/Publications/ziska/potentialthreats.pdf .

Ziska, L. H., et al. Rising CO_2, climate change, and public health: Exploring the links to plant biology. *Environmental Health Perspectives* 117(2):155–158, 2009.

CHAPTER 3: THE HOT ZONE

Brown, R. Summer's extreme weather. *Salon*, August 20, 2010. www.salon.com/2010/08/20/extreme_weather_slide_show.

Crumley, B. Elder careless. *Time*, August 24, 2003. www.time.com/time/magazine/article/0,9171,477899,00.html.

Dear, K., et al. Effects of temperature and ozone on daily mortality during the August 2003 heat wave in France. *Archives of Environmental and Occupational Health* 60(4):205–212, 2005.

Dematte, J. E., et al. Near-fatal heat stroke during the 1995 heat wave in Chicago. *Annals of Internal Medicine* 129(3):173–181, 1998.

Deppe, B. Arizona faces largest wildfire in its history. *People's World*, June 17, 2011. www.peoplesworld.org/arizona-faces-largest-wildfire-in-its-history.

Editorial Staff. Phoenix's urban heat island. *Phoenix Magazine*, May 2011. www.phoenixmag.com/lifestyle/valley-news/201105/phoenix-s-urban-heat-island.

Evans, J. Priority one: Protect the environment jobs. *Grand Junction Free Press*, March 4, 2011. www.gjfreepress.com/article/20110304/COLUMNISTS/110309968.

Freedman, A. Heat wave sets more all-time temperature records. Extreme Planet [blog], ClimateCentral.org, July 1, 2012. www.climatecentral.org/blogs/heat-wave-sets-more-all-time-temperature-records.

Gardner, E. K., and Kline, G. Researchers find future temperatures could exceed livable limits. Purdue University News Service, May 4, 2012. www.purdue.edu/newsroom/research/2010/100504HuberLimits.html.

Ioffe, J. Russia on fire. News Desk [blog], NewYorker.com, August 5, 2010. www.newyorker.com/online/blogs/newsdesk/2010/08/russia-fires.html.

Isachenkov, V. Moscow's lethal smog. *Salon*, August 9, 2010. www.stage1.salon.com/2010/08/09/eu_russia_fires_1.

Kalkstein, L., et al. Chapter 4: Health impacts of heat: Present realities and potential impacts of a climate change. *Impacts of climate change and disasters: Concepts and cases*, Matthias, R., and Ibarrarán, M. E., eds. New Horizons in Environmental Economics. Cheltenham, UK: Edward Elgar, 2009. www.as.miami.edu/geography/research/climatology/Chapter_04.pdf.

Kalkstein, L. S., et al. Analog European heat waves for U.S. cities to analyze impacts on heat-related mortality. *Bulletin of the American Meteorological Society* 89:75–85, 2008. http://journals.ametsoc.org/doi/pdf/10.1175/BAMS-89-1-75.

Karl, T. R., and Knight, R. W. The 1995 Chicago heat wave: How likely is a recurrence? *Bulletin of the American Meteorological Society* 78:1107–1119, 1997. http://dx.doi.org/10.1175/1520-0477(1997)078<1107:TCHWH L>2.0.CO;2.

Kaufman, L. A city prepares for a warm long-term forecast. *New York Times*, May 22, 2011. www.nytimes.com/2011/05/23/science/earth/23adaptation.html.

Khamsi, R. Human activity implicated in Europe's 2003 heat wave. Nature.com, December 1, 2004. www.nature.com/news/2004/041129/full/news041129-6.html.

Klinenberg, E. Dead heat. *Slate*, July 30, 2002. www.slate.com/articles/health_and_science/medical_examiner/2002/07/dead_heat.html.

———. When Chicago baked. *Slate*, September 2, 2005. www.slate.com/articles/news_and_politics/history_lesson/2005/09/when_chicago_baked.html.

Kovats, R. S., and Kristie, L. E. Heatwaves and public health in Europe. *European Journal of Public Health* 16(6):592–599, 2006. http://eurpub.oxfordjournals.org/content/16/6/592.long.

Kramer, A. E. Russia, crippled by drought, bans grain exports. *New York Times*, August 5, 2010. www.nytimes.com/2010/08/06/world/europe/06russia.html.

Landau, E. Heatstroke: A deadly hazard of summer. CNN.com, July 14, 2011. www.cnn.com/2011/HEALTH/07/13/heat.illnesses.heatstroke/index.html.

Loiko, S. L. Moscow death toll soars as heat wave persists. *Los Angeles Times*, August 10, 2010. http://articles.latimes.com/2010/aug/10/world/la-fg-russia-heat-deaths-20100810.

Luber, G., and Hess, J. Climate change and human health in the United States. *Journal of Environmental Health* 70(5): 43–46, 2007.

Luber, G., and McGeehin, M. Climate change and extreme heat events. *American Journal of Preventive Medicine* 35(5):429–435, 2008. www.ajpmonline.org/article/S0749-3797(08)00686-7/fulltext.

Managing Director's Office of Emergency Management. Excessive heat plan. City of Philadelphia, n.d. http://oem.readyphiladelphia.org/HeatPlan.

Martelle, S. Russian drought spurs worldwide food price hikes. *AOL News*, September 1, 2010. www.aolnews.com/2010/09/01/russian-drought-spurs-worldwide-food-price-hikes.

McElroy, D. Russian heatwave kills 5,000 as fires rage out of control. *The Telegraph*, August 6, 2010. www.telegraph.co.uk/news/worldnews/europe/russia/7931206/Russian-heatwave-kills-5000-as-fires-rage-out-of-control.html.

Mechanic, M. And you thought that heat wave was bad? MotherJones.com, March 28, 2012. www.motherjones.com/environment/2012/03/heat-wave-climate-change-future-matthew-huber-interview.

Minters, B. Too darn hot…but not so deadly. *Philadelphia Inquirer*, August 30, 2010. http://articles.philly.com/2010-08-30/news/24973855_1_heat-wave-heat-warning-heat-hits.

National Oceanic and Atmospheric Administration. *Natural disaster survey report: July 1995 heat wave*. Silver Spring, MD: National Weather Service, 1995. www.nws.noaa.gov/om/assessments/pdfs/heat95.pdf.

Natural Resources Defense Council. Extreme heat: More intense hot days and heat waves. NRDC.org, n.d. www.nrdc.org/health/climate/heat.asp.

Ohlheiser, A., et al. Power outages dampen July 4 parties. *Slate*, July 4, 2012. http://slatest.slate.com/posts/2012/06/30/washington_d_c_maryland_virginia_derecho_storm_leaves_2_mil_without_power_kills_4_.html.

Portier, C. J., et al. *A human health perspective on climate change*. Research Triangle Park, NC: Environmental Health Perspectives/National Institute of Environmental Health Sciences, 2010. www.niehs.nih.gov/health/assets/docs_a_e/climatereport2010.pdf.

Poumadère, M., et al. The 2003 heat wave in France: Dangerous climate change here and now. *Risk Analysis* 25(6):1483–1494, 2005.

Quinton, S. As politicians debate climate change, our forests wither. TheAtlantic.com, June 15, 2012. www .theatlantic.com/national/archive/2012/06/as-politicians-debate-climate-change-our-forests-wither/258549.

Rappold, A. G., et al. Peat bog wildfire smoke exposure in rural North Carolina is associated with cardiopul-monary emergency department visits assessed through syndromic surveillance. *Environmental Health Per-spectives* 119(10):1415–1420, 2011. www.ncbi.nlm.nih.gov/pmc/articles/PMC3230437.

Rice, D. Wildfires and drought plague much of the southern USA. *USA Today*, June 10, 2012. www.usatoday.com/weather/drought/2011-06-09-drought-wildfires-Arizona-Wallow-Fire_n.htm.

Romm, J. "Global warming is a medical emergency": Hellish heatwaves to harm health of millions. Think-Progress.org, August 26, 2009. http://thinkprogress.org/climate/2009/08/26/204550/global-warming-health-impacts-heat-waves-ps.

Schiffman, R. We started the fire: Global warming burns the West. *Salon*, July 3, 2012. www.salon.com/ 2012/07/03/we_started_the_fire_global_warming_burns_the_southwest.

Semenza, J. C., et al. Heat-related deaths during the July 1995 heat wave in Chicago. *New England Journal of Medicine* 335(2):84–90, 1996. www.nejm.org/doi/full/10.1056/NEJM199607113350203.

Sheridan, S. C., and Kalkstein, L. S. Progress in heat-watch warning system technology. *Bulletin of the American Meteorological Society* 85(12):1931–1941, 2004. http://sheridan.geog.kent.edu/pubs/ 2004-BAMS.pdf.

Sheridan, S. C., et al. Trends in heat-related mortality in the United States, 1975–2004. *Natural Hazards* 50:145–160, 2009. http://sheridan.geog.kent.edu/pubs/2008-NH.pdf.

Sherwood, S. C., and Huber, M. An adaptability limit to climate change due to heat stress. *Proceedings of the National Academy of Sciences of the United States of America* 107(21):9552–9555, 2010. www.ncbi.nlm.nih. gov/pmc/articles/PMC2906879.

Shuster, S. Russia still dragging its feet on climate change. Time.com, November 4, 2009. www.time.com/ time/specials/packages/article/0,28804,1929071_1929070_1934785,00.html.

Shwartz, M. Heat waves and extremely high temperatures could be commonplace in the U.S. by 2039, Stan-ford study finds. Stanford News, July 8, 2010. http://news.stanford.edu/news/2010/july/extreme-heat-study-070810.html.

Stott, P. A., et al. Human contribution to the European heatwave of 2003. *Nature* 432:610–614, 2004. www.csag.uct.ac.za/~daithi/papers/StottPA_StoneDA_AllenMR_2004.pdf.

Study: Climate change to increase Yellowstone wildfires dramatically. University of California Merced, Uni-versity News, July 25, 2011. www.ucmerced.edu/news/study-climate-change-increase-yellowstone-wildfires-dramatically.

Sun, L. H. Medical examiners use differing criteria to tally heat-related deaths. *Washington Post*, July 27, 2010. www.washingtonpost.com/wp-dyn/content/article/2010/07/26/AR2010072604723.html.

Taylor, C. Acceptable losses. *Salon*, August 20, 2002. www.salon.com/2002/08/20/heat.

Thean, T. Wildfires: They're not just dangerous to trees. Ecocentric [blog], Time.com, June 29, 2011. http:// ecocentric.blogs.time.com/2011/06/29/wildfires-theyre-not-just-dangerous-to-trees.

Tucker, C. Russian government sees harsh evidence of global warming. Cynthia Tucker [blog], AJC.com, August 16, 2012. http://blogs.ajc.com/cynthia-tucker/2010/08/16/russian-government-sees-harsh-evidence-of-global-warming.

United States Environmental Protection Agency Office of Atmospheric Programs. *Excessive heat events guidebook*. EPA 430-B-06-005, June 2006. Washington, DC: US Environmental Protection Agency, 2006. www.epa.gov/hiri/about/pdf/EHEguide_final.pdf.

University Corporation for Atmospheric Research. U.S. fires release large amounts of carbon dioxide, new study shows [press release]. October 31, 2007. www.ucar.edu/news/releases/2007/co2fires.jsp.

University of Chicago Medicine. Classic heat stroke during Chicago 1995 heat wave [press release]. August 1, 1998. www.uchospitals.edu/news/1998/19980801-heatstroke-aim.html.

Walker, S. Russian wildfires: "Even the road seemed to be on fire. It was like descending into hell." *The Inde-pendent*, August 5, 2010. www.independent.co.uk/news/world/europe/russian-wildfires-even-the-road-seemed-to-be-on-fire-it-was-like-descending-into-hell-2043553.html.

Weinmann, A. More extreme heat waves: Global warming's wake-up call [press release]. NWF.org, September 18, 2010. www.nwf.org/News-and-Magazines/Media-Center/Reports/Archive/2009/Heat-Waves.aspx.

Welch, C. Climate change, beetle may doom rugged pine. *Seattle Times*, November 5, 2011. http://seattletimes.nwsource.com/html/localnews/2016699269_barkbeetle06m.html.

Westerling A. L., et al. Warming and earlier spring increase western U.S. forest wildfire activity. *Science* 313:940–943, 2006.

Whitman, S., et al. Mortality in Chicago attributed to the July 1995 heat wave. *American Journal of Public Health* 87(9):1515–1518, 1997. www.ncbi.nlm.nih.gov/pmc/articles/PMC1380980/pdf/amjph00508-0117.pdf.

Williams, S. Residents flee as acrid smog blankets Moscow. *Sydney Morning Herald*, August 7, 2010. http://news.smh.com.au/breaking-news-world/residents-flee-as-acrid-smog-blankets-moscow-20100807-11p7z.html.

World Health Organization Regional Office for Europe. *Euroheat: Improving public health responses to extreme weather/heat-waves: Summary for policy-makers.* Copenhagen: World Health Organization, 2009. www.euro.who.int/en/what-we-do/health-topics/environment-and-health/Climate-change/publications/2009/euroheat.-improving-public-health-responses-to-extreme-weatherheat-waves.

———. Extreme heat-wave and wildfires cause public health concerns in the Russian Federation. World Health Organization, August 17, 2010. www.euro.who.int/en/what-we-do/health-topics/environment-and-health/air-quality/news/news/2010/08/extreme-heat-wave-and-wildfires-cause-public-health-concerns-in-the-russian-federation.

———. Wild fires and heat-wave in the Russian Federation. Situation Report, August 10, 2010. www.euro.who.int/__data/assets/pdf_file/0010/119764/wildfires_rus_sitrep100810.pdf.

CHAPTER 4: HEALTH CARE ON LIFE SUPPORT

In addition to the sources mentioned in the text, Dr. Anthony Speier, assistant secretary for the Office of Behavioral Health in the Louisiana Department of Health and Hospitals, gave generously of his time to bring me up to speed on the serious and lingering psychological damage sustained by Katrina survivors, many of whom suffered from PTSD and severe depression. Bethany Bultman, cofounder, president, and director of the New Orleans Musicians Clinic, helped illuminate in real-life terms the toll extracted by the collapse of the health care system after Katrina and how musicians and other members of New Orleans's creative community have struggled to survive.

Berggren, R. Unexpected necessities—Inside Charity Hospital. *New England Journal of Medicine* 353(15):1550–1553. www.nejm.org/doi/full/10.1056/NEJMp058239.

Berggren, R. E., and Curiel, T. J. After the storm—Health care infrastructure in post-Katrina New Orleans. *New England Journal of Medicine* 354(15):1549–1552, 2006. www.nejm.org/doi/full/10.1056/NEJMp068039.

Carreau, M. How did modest Katrina morph into monster? *Houston Chronicle*, September 12, 2005. www.chron.com/news/hurricanes/article/How-did-modest-Katrina-morph-into-monster-1483672.php.

Chen, P. W. Tending to patients during a hurricane. *New York Times*, September 2, 2010. www.nytimes.com/2010/09/02/health/views/02chen.html.

Commonwealth Fund. New report: Post-Katrina safety-net clinic patients in New Orleans report more efficient and affordable health care and less medical debt than most U.S. adults; pilot could serve as national model [press release]. January 15, 2010. www.commonwealthfund.org/News/News-Releases/2010/Jan/Post-Katrina-Safety-Net-Clinic-Patients.aspx.

Connolly, C. New Orleans health care another Katrina casualty. *Washington Post*, November 25, 2005. www.washingtonpost.com/wp-dyn/content/article/2005/11/24/AR2005112400730.html.

deBoisblanc, B. P. Black Hawk, please come down: Reflections on a hospital's struggle to survive in the wake of Hurricane Katrina. *American Journal of Respiratory and Critical Care Medicine* 172(10):1239–1240, 2005. http://ajrccm.atsjournals.org/content/172/10/1239.

deBoisblanc, B. P., and Kantrow, S. P. A letter from New Orleans five years later. *American Journal of Respiratory and Critical Care Medicine* 182(8):989–990, 2010. http://ajrccm.atsjournals.org/content/182/8/989.

Deichmann, R. E. *Code blue: A Katrina physician's memoir.* Bloomington, IN: Rooftop Publishing, 2007.

DeSalvo, K., et al. The nine o'clock meeting. *Health Affairs* 25(2):483, 2006. http://content.healthaffairs.org/content/25/2/483.1.full.

Editors. Debunking the myths of Hurricane Katrina: Special report. *Popular Mechanics,* February 15, 2006. www.popularmechanics.com/science/environment/natural-disasters/2315076.

Harrison, E. Suffering a slow recovery. *Scientific American,* December 2007. www.scientificamerican.com/article.cfm?id=katrina-suffering-a-slow-recovery.

Hurricane Katrina increased mental and physical health problems in New Orleans by up to three times. *Science Daily,* September 3, 2008. www.sciencedaily.com/releases/2008/09/080903075614.htm.

Lamberg, L. Katrina survivors strive to reclaim their lives. *JAMA: The Journal of the American Medical Association* 296(5):499–502, 2006.

Liu, A., et al., eds. *Resilience and opportunity: Lessons from the U.S. Gulf Coast after Katrina and Rita.* Washington DC: Brookings Institution, 2011.

Louisiana State University. LSU is site of largest acute-care field hospital in U.S. history [press release]. September 6, 2005. www.lsu.edu/pa/mediacenter/nr/20050906_645p.htm.

McCarthy, M. New Orleans struggles to rebuild its health system. *The Lancet* 368(9541):1056—1058, 2006. www.thelancet.com/journals/lancet/article/PIIS0140-6736(06)69427-0/fulltext.

Nossiter, A. Plan for New Orleans hospitals draws outcry. *New York Times,* November 25, 2008. www.nytimes.com/2008/11/26/us/26hospital.html.

O'Reilly, K. B. Katrina's legacy: Moving beyond the storm. *American Medical News,* August 16, 2010. www.ama-assn.org/amednews/2010/08/16/prsa0816.htm.

Rich, N. Jungleland. *New York Times Magazine,* March 21, 2012. www.nytimes.com/2012/03/25/magazine/the-lower-ninth-ward-new-orleans.html.

Schiermeier, Q. Climate and weather: Extreme measures. *Nature* 477:148–149, 2011. www.nature.com/news/2011/110907/full/477148a.html.

Schwartz, J. Two studies link global warming to greater power of hurricanes. *New York Times,* May 31, 2006. www.nytimes.com/2006/05/31/science/31climate.html.

Shute, N. On life support. *US News and World Report,* April 16, 2006. http://health.usnews.com/usnews/health/articles/060424/24charity.htm.

Tidwell, M. *The ravaging tide: Strange weather, future Katrinas, and the coming death of America's coastal cities.* New York: Free Press, 2006.

Treaster, J. B. Superdome: Haven quickly becomes an ordeal. *New York Times,* September 1, 2005. www.nytimes.com/2005/09/01/national/nationalspecial/01dome.html.

Voelker, R. In post-Katrina New Orleans, efforts under way to build better health care. *JAMA: The Journal of the American Medical Association* 296(11):1333–1334, 2006.

Zuckerman, S., and Coughlin, T. A. *Initial health policy responses to Hurricane Katrina and possible next steps.* Washington, DC: Urban Institute, February 2006. www.urban.org/publications/900929.html.

Chapter 5: Running on Empty

Barry Nelson, a senior policy analyst for NRDC's water program in San Francisco, helped me sort through many of the complex and conflicting environmental issues involved in the battles over water in California's Sacramento Delta. Conversations with former Fresno Bee reporter and environmental activist Lloyd Carter gave me an in-depth sense of the history and the political jockeying that goes on behind closed doors regarding allocations of this precious resource.

These materials were also invaluable:

5 questions for geologist Jeff Mount on California's crumbling delta levees. PopularMechanics.com, October 1, 2009. www.popularmechanics.com/science/4258291.

Amato, I. Genes take a back seat. *Chemical and Engineering News* 87(14):28–32, 2009. https://pubs.acs.org/cen/science/87/8714sci1.html.

Barnett, T. P., and Pierce, D. W. When will Lake Mead go dry? *Water Resources Research* 44(3):W3201, 2008. www.image.ucar.edu/idag/Papers/PapersIDAGsubtask2.4/Barnett1.pdf.

Barringer, F. Las Vegas's worried water czar. Green [blog], NYTimes.com, September 28, 2010. http://green.blogs.nytimes.com/2010/09/28/las-vegass-worried-water-czar.

———. Water use in Southwest heads for a day of reckoning. *New York Times*, September 27, 2010. www.nytimes.com/2010/09/28/us/28mead.html.

Basu, M. Drought stretches across America, threatens crops. CNN.com, July 13, 2012. www.cnn.com/2012/07/13/us/midwest-drought/index.html.

Battisti, D. S., and Naylor, R. L. Historical warnings of future food insecurity with unprecedented seasonal heat. *Science* 323(5911):240–244, 2009. http://people.fas.harvard.edu/~eebutler//Homepage/2010-2011_files/battisti_naylor_2009.pdf.

Biello, D. Can climate change cause conflict? Recent history suggests so. ScientificAmerican.com, November 23, 2008. www.scientificamerican.com/article.cfm?id=can-climate-change-cause-conflict.

Bierbaum, R. M., et al., eds. *Confronting climate change: Avoiding the unmanageable and managing the unavoidable.* Research Triangle Park, NC: Sigma Xi, and Washington, DC: United Nations Foundation, 2007. www.sigmaxi.org/programs/unseg/Full_Report.pdf.

Burke, G. Calif. river system is nation's most endangered. *The Guardian*, April 7, 2009. www.guardian.co.uk/world/feedarticle/8443035.

Burke, M. B., et al. Warming increases the risk of civil war in Africa. *Proceedings of the National Academy of Sciences of the United States of America* 106(49):20670–20674, 2009. www.pnas.org/content/106/49/20670.full.

Carey, N. *The epigenetics revolution: How modern biology is rewriting our understanding of genetics, disease and inheritance.* New York: Columbia University Press, 2012.

Craft, B. The men who made Las Vegas: Bugsy and the mob. *Strip Las Vegas*, Issue 57, 2011. www.striplv.com/site/mag/issue57/stories/Vegas-BugsySiegel.html.

Dai, A. Drought under global warming: A review. *Wiley Interdisciplinary Reviews: Climate Change* 2(1):45–65, 2011.

Deschênes, O., and Moretti, E. Extreme weather events, mortality, and migration. *Review of Economics and Statistics* 91(4):659–681, 2009.

Eligon, J. Drought disrupts everyday tasks in rural Midwest. *New York Times*, August 23, 2012. www.nytimes.com/2012/08/24/us/midwest-water-wells-drying-up-in-drought.html.

Elliott, L. World Bank issues hunger warning after droughts in US and Europe. *The Guardian*, August 30, 2012. www.guardian.co.uk/global-development/2012/aug/30/world-bank-global-hunger-warning-drought.

Gillis, J. A warming planet struggles to feed itself. *New York Times*, June 4, 2011. www.nytimes.com/2011/06/05/science/earth/05harvest.html.

Gleick, P. Why don't we get our drinking water from the ocean by taking the salt out of seawater? ScientificAmerican.com, July 23, 2008. www.scientificamerican.com/article.cfm?id=why-dont-we-get-our-drinking-water-from-the-ocean.

Green, E. Quenching Las Vegas' Thirst: Part 1: Satiating a booming city. *Las Vegas Sun*, June 1, 2008. www.lasvegassun.com/news/2008/jun/01/satiating-booming-city.

Green, E. Quenching Las Vegas' Thirst: Part 2: The chosen one. *Las Vegas Sun*, June 8, 2008. www.lasvegassun.com/news/2008/jun/08/chosen-one.

Green, E. Quenching Las Vegas' Thirst: Part 3: The equation: No water, no growth. *Las Vegas Sun*, June 15, 2008. www.lasvegassun.com/news/2008/jun/15/equation-no-water-no-growth.

Jaffe, A. Lake Mead, the American Southwest, and water: An interview with Tim Barnett. CircleofBlue.org, July 8, 2008. www.circleofblue.org/waternews/2008/world/north-america/lake-mead-the-american-southwest-and-water-an-interview-with-tim-barnett.

Jirtle, R. L., and Skinner, M. K. Environmental epigenomics and disease susceptibility. *Nature Reviews Genetics* 8(4):253–262, 2007.

Kolbert, E. The big fix. *OnEarth*, Fall 2008. www.onearth.org/article/the-big-fix.

———. Comment: The big heat. *New Yorker*, July 23, 2012. www.newyorker.com/talk/comment/2012/07/23/120723taco_talk_kolbert.

Long, S. P., ed. Virtual special issue on food security—Greater than anticipated impacts of near-term global atmospheric change on rice and wheat [editorial]. *Global Change Biology* 18(5):1489–1490.

Lowrey, A. Experts issue a warning as food prices shoot up. *New York Times,* September 4, 2012. www.nytimes.com/2012/09/05/business/experts-issue-a-warning-as-food-prices-shoot-up.html.

———. Protesting on an empty stomach. *Slate*, January 31, 2011. www.slate.com/articles/business/moneybox/2011/01/protesting_on_an_empty_stomach.html.

Lowrey, A., and Nixon, R. Severe drought seen as driving cost of food up. *New York Times*, July 25, 2012. www.nytimes.com/2012/07/26/business/food-prices-to-rise-in-wake-of-severe-drought.html.

Lumey, L. H., et al. Cohort profile: The Dutch Hunger Winter Families study. *International Journal of Epidemiology* 36(6):1196–1204, 2007. http://ije.oxfordjournals.org/content/36/6/1196.full.

Mulroy, P. Where I stand—Guest Column: Stretching the Colorado River. *Las Vegas Sun*, August 12, 2012. www.lasvegassun.com/news/2012/aug/12/stretching-colorado-river.

Munoz, D. Australia floods cause "catastrophic" damage. Reuters.com, January 5, 2011. www.reuters.com/article/2011/01/05/us-australia-floods-idUSTRE6BU09620110105.

Muskal, M. As drought widens, 50.3% of U.S. counties declared disaster areas. *Los Angeles Times*, August 1, 2012. http://articles.latimes.com/2012/aug/01/nation/la-na-nn-drought-strikes-over-half-of-us-20120801.

Naylor, R., and Falcon, W. Our daily bread. *Boston Review*, September/October 2008. http://bostonreview.net/BR33.5/naylorfalcon.php.

Nelson, B. Delta Stewardship Council keys to success—Science and adaptive management. Barry Nelson [blog], Switchboard.nrdc.org, April 29, 2010. http://switchboard.nrdc.org/blogs/bnelson/delta_stewardship_council_keys_2.html.

Nelson, B., et al. *In Hot Water: Water Management Strategies to Weather the Effects of Global Warming.* Natural Resources Defense Council, July 2007. www.nrdc.org/globalwarming/hotwater/hotwater.pdf.

O'Neill, G. Feature: Designing drought tolerant crops. *Life Scientist*, November 15, 2010. www.lifescientist.com.au/article/368007/feature_designing_drought_tolerant_crops.

Power, M. Peak water: Aquifers and rivers are running dry. How three regions are coping. *Wired*, Issue 16.05, May 2008. www.wired.com/science/planetearth/magazine/16-05/ff_peakwater.

Quinlan, P. Lake Mead's water level plunges as 11-year drought lingers. *New York Times*, August 13, 2010. www.nytimes.com/gwire/2010/08/12/12greenwire-lake-meads-water-level-plunges-as-11-year-drou-29594.html.

Schlenker, W., and Roberts, M. J. Nonlinear temperature effects indicate severe damages to U.S. crop yields under climate change. *Proceedings of the National Academy of Sciences of the United States of America* 106(37):15594–15598, 2009. www.pnas.org/content/106/37/15594.full.

Severson, K., and Johnson, K. Drought spreads pain from Florida to Arizona. *New York Times*, July 11, 2011. www.nytimes.com/2011/07/12/us/12drought.html.

Song, S., et al. Famine, death, and madness: Schizophrenia in early adulthood after prenatal exposure to the Chinese Great Leap Forward Famine. *Social Science and Medicine* 68(7):1315–1321, 2009.

Specter, M. The last drop. *New Yorker*, October 23, 2006. www.newyorker.com/archive/2006/10/23/061023fa_fact1.

Walsh, B. When the rains stop. *Time*, August 6, 2012. www.time.com/time/magazine/article/0,9171,2120491,00.html.

Watters, E. DNA is not destiny: The new science of epigenetics rewrites the rules of disease, heredity, and identity. *Discover*, November 2006. http://discovermagazine.com/2006/nov/cover.

Yang, S. Can California fix the Delta before disaster strikes? University of California at Berkeley News Center, April 20, 2010. http://newscenter.berkeley.edu/2010/04/20/delta.

Zimmerman, J. Sacramento–San Joaquin Delta becomes water war's front line. *Press-Enterprise*, March 21, 2009. www.pe.com/local-news/reports/water/water-headlines/20090321-sacramento-san-joaquin-delta-becomes-water-wars-front-line.ece.

CHAPTER 6: THROUGH A GLASS DARKLY

In addition to the sources mentioned in the text, Australian journalists Fiona Armstrong and Melissa Sweet gave generously of their time to provide me with perspective and introduce me to key sources. Dr. Desley Marshall and her husband, Robert Buchan, were incredibly gracious and showed me around St. George and made introductions to local residents so I could get a real feel of what happened during the 2011 floods. I could not have done any of this without the help of Marg Moss, executive director of the Rural Doctors Association of Queensland in Brisbane, who spent many hours setting up interviews for me with member physicians throughout the hard hit agricultural areas in eastern Australia.

Other key background materials:

Australian Government. National Water Commission annual report 2009–10. Commonwealth of Australia, 2010. http://archive.nwc.gov.au/__data/assets/pdf_file/0004/7528/Annual_Report_2009-10.pdf.

Brisbane under water: Your stories. BBC News, January 12, 2011. www.bbc.co.uk/news/world-asia-pacific-12169592.

Cameron, P. A., et al. Black Saturday: The immediate impact of the February 2009 bushfires in Victoria, Australia. *Medical Journal of Australia* 191(1):11–16, 2009. https://www.mja.com.au/journal/2009/191/1/black-saturday-immediate-impact-february-2009-bushfires-victoria-australia.

Cleugh, H., et al., eds. *Climate change: Science and solutions for Australia*. Collingwood, Australia: CSIRO, 2011. www.publish.csiro.au/pid/6558.htm.

Climate Institute. Factsheet: *Climate change and extreme weather events. February 2011*. www.climateinstitute.org.au/verve/_resources/tci_extremeweatherfactsheet_feb2011.pdf.

CSIRO. +4C scenarios for Australia's future climate. Phys.org, July 12, 2011. http://phys.org/news/2011-07-4c-scenarios-australia-future-climate.html.

———. Adapting agriculture to climate change. Reference 07/240. December 4, 2007. www.csiro.au/Organisation-Structure/Divisions/Ecosystem-Sciences/AdaptingAgriculture.aspx.

———. *The science of providing water solutions for Australia*. Melbourne: CSIRO, 2010.

Draper, R. Australia's dry run. *National Geographic*, April 2009. http://ngm.nationalgeographic.com/print/2009/04/murray-darling/draper-text.

Fenner, R., and Daley, G. Cyclone Yasi, stronger than Katrina, hits Australia. Bloomberg.com, February 2, 2011. www.bloomberg.com/news/2011-02-02/cyclone-yasi-slams-australian-coast-bringing-winds-stronger-than-katrina.html.

Fogarty, D. Scientists see climate change link to Australian floods. Reuters.com, January 12, 2011. www.reuters.com/article/2011/01/12/us-climate-australia-floods-idUSTRE70B1XF20110112.

Friedman, T. Elephants down under [op-ed]. *New York Times*, March 27, 2012. www.nytimes.com/2012/03/28/opinion/friedman-elephants-down-under.html.

Fritze, J. G., et al. Hope, despair and transformation: Climate change and the promotion of mental health and wellbeing. *International Journal of Mental Health Systems* 2:13, 2008. www.ncbi.nlm.nih.gov/pmc/articles/PMC2556310.

Garnaut, R. *The Garnaut climate change review: Final report*. Australian Government, 2008. www.garnautreview.org.au/index.htm.

———. *The Garnaut climate change review—Update 2011*. Australian Government, 2011. www.garnautreview.org.au/update-2011/garnaut-review-2011.html.

Goodell, J. Climate change and the end of Australia. *Rolling Stone*, October 3, 2011. www.rollingstone.com/politics/news/climate-change-and-the-end-of-australia-20111003.

Hughes, Lesley, and Tony McMichael. *The Critical Decade: Climate Change and Health* (Climate Commission; Nov. 2011) http://climatecommission.gov.au/wp-content/uploads/111129_FINAL-FOR-WEB.pdf

Morgan, G. Lessons from the ashes. *Wildfire Magazine*, September/October 2009. http://wildfiremag.com/wui/black-saturday-bushfire-research-200909.

Morton, A. Extreme weather is just the beginning: Garnaut. *The Age*, February 4, 2011. www.theage.com.au/national/extreme-weather-is-just-the-beginning-garnaut-20110203-1afgr.html.

Newman, E. Global boiling: Australia's Black Saturday of extreme fire. ThinkProgress.org, March 1, 2009. http://thinkprogress.org/climate/2009/03/01/174269/black-saturday-fire.

Oakeshott, G., and Maslen, G. The farmer wants a future. *Agribusiness Chain* 10(1):15–24, 2010. www.scribd.com/doc/35442092/Agribusiness-v10-1.

Pugh, W. Australian farms 'vulnerable' to climate change, adviser Garnaut says. Bloomberg.com, March 2, 2011. www.bloomberg.com/news/2011-03-02/australian-farms-vulnerable-to-climate-change-garnaut-says.html.

Strudwick, D. 30,000 ordered to evacuate as Cairns prepares for Cyclone Yasi. Cairns.com.au, February 1, 2011. www.cairns.com.au/article/2011/02/01/147281_cyclone.html.

Wilkinson, M., and Cubby, B. The end of climate certainty. *Sydney Morning Herald*, February 14, 2009. www.smh.com.au/environment/climate-change/the-end-of-climate-certainty-20090213-876f.html.

Windram, C. Drought in Australia—The lessons we can learn for tackling climate change. Think Carbon [blog]. June 22, 2009. http://thinkcarbon.wordpress.com/2009/06/22/drought-in-australia---the-lessons-we-can-learn-for-tackling-climate-change.

Yun Low, W. (2011). Asia-Pacific Journal of Public Health: Supplement issue on climate change. *Asia-Pacific Journal of Public Health*, 23(2).

CHAPTER 7: HOLDING BACK THE WATERS

In addition to the sources cited in the chapter, Sandy Rosenthal of Levees.org in New Orleans was especially helpful in discussing the ongoing damage to the levees. Water engineer Peter Binney of Merrick & Company generously spent time educating me about what needs to be done to create a sustainable infrastructure and how to find new water sources in the Southwest. David Sedlak, codirector of Berkeley Water Center and director of the Institute for Environmental Science and Engineering at UC Berkeley, outlined what cities need to do, now and in the future, to avoid running dry. Barb Graff, director of Emergency Management for the City of Seattle, and Jill Simmons, director of the Office of Sustainability and the Environment, generously provided me with information on Seattle's innovative climate adaptation plans and the preparations the region is making to limit the damage of future weather-induced disasters.

Key written materials:

Barone, J. From toilet to tap. *Discover*, May 2008. http://discovermagazine.com/2008/may/23-from-toilet-to-tap.

Barry, J. M. Why do we leave New Orleans dangerously exposed to the next big storm? *The Daily Beast*, August 30, 2012. www.thedailybeast.com/articles/2012/08/28/why-do-we-leave-new-orleans-dangerously-exposed-to-the-next-big-storm.html.

Blake, S. Miami Beach wades into sea level rise. *Miami Today*, August 23, 2012. http://miamitodaynews.com/news/120823/story2.shtml.

Bourne, J. Plumbing California: California's pipe dream. *National Geographic*, April 2010. http://ngm.nationalgeographic.com/2010/04/plumbing-california/bourne-text.

Bowser, B. A. Will new levees protect New Orleans from the next hurricane? The Rundown [blog], PBS News-hour, August 26, 2010. www.pbs.org/newshour/rundown/2010/08/five-years-after-katrina-some-question-whether-new-levees-will-protect-the-city-next-time.html.

Davis, T. Underwater cities: Climate change begins to reshape the urban landscape. *Grist*, October 27, 2011. http://grist.org/cities/2011-10-26-underwater-cities-climate-change-begins-reshape-urban-landscape.

Fischetti, M. Drowning New Orleans. *Scientific American*, October 2001. www.scientificamerican.com/article.cfm?id=drowning-new-orleans-hurricane-prediction.

———. New Orleans protection plan will rely on wetlands to hold back hurricanes. Observations [blog], ScientificAmerican.com, January 26, 2012. http://blogs.scientificamerican.com/observations/2012/01/26/new-orleans-protection-plan-will-rely-on-wetlands-to-hold-back-hurricanes.

Galloway, G. E. A plea for a coordinated national water policy. *Bridge* 41(4):37–46, Winter 2011. www.nae.edu/File.aspx?id=55285.

Jefferson, A. Levees and the illusion of flood control. Highly Allochthonous [blog], All-Geo.org, May 19, 2011. http://all-geo.org/highlyallochthonous/2011/05/levees-and-the-illusion-of-flood-control.

Kix, P. You are drinking what? *Wall Street Journal*, August 24, 2012. http://online.wsj.com/article/SB10000872396390444270404577607333668861496.html.

Lehmann, E. When the levee breaks: U.S. flood protection inadequate. ScientificAmerican.com, January 18, 2012. www.scientificamerican.com/article.cfm?id=us-flood-protection-inadequate-levee-breaks.

Martin, T. W., and Esterl, M. Reckoning with Isaac. *Wall Street Journal*, August 30, 2012. http://online.wsj.com/article/SB10000872396390443618604577621212032412488.html.

Miami-Dade County. Climate change action plan. In *GreenPrint: Our design for a sustainable future*. December 2010. www.miamidade.gov/greenprint/pdf/plan.pdf.

Morgan, C. Rising sea comes at a cost for South Florida cities. *Miami Herald*, September 1, 2012. www.miamiherald.com/2012/09/01/2980388/rising-sea-come-at-a-cost-for.html.

Padgett, T. Hard times in the Big Easy: Restoring wetlands key to avoiding another Katrina. *Time*, August 27, 2010. www.time.com/time/specials/packages/article/0,28804,2012217_2012252_2014247,00.html.

Schleifstein, M. Levee statistics point up their importance to nation's economy. *Times-Picayune*, January 2, 2010. www.nola.com/hurricane/index.ssf/2010/01/levee_statistics_point_up_thei.html.

———. New Orleans levees get a near-failing grade in new Corps rating system. *Times-Picayune*, August 29, 2011. www.nola.com/environment/index.ssf/2011/08/new_orleans_levees_get_a_near-.html.

Schwartz, J. Vast defenses now shielding New Orleans. *New York Times*, June 14, 2012. www.nytimes.com/2012/06/15/us/vast-defenses-now-shielding-new-orleans.html.

US Military, Army. (2012). *MRGO ecosystem restoration*. Retrieved from U.S. Army Corps of Engineers. www.mrgo.gov/MRGO_restoration_study.aspx

Wolman, D. Before the levees break: A plan to save the Netherlands. *Wired*, Issue 17.01, January 2009. www.wired.com/science/planetearth/magazine/17-01/ff_dutch_delta.

CHAPTER 8: THE EMERALD CITY

Canadian urban planner and landscape architect Mark Holland, who founded the City of Vancouver's sustainability office and helped transform False Creek into the livable community it is today, provided valuable insights into how Vancouver became a model of sustainability. I also benefited from discussions with theoretical physicist Geoffrey West of the Santa Fe Institute in New Mexico about the fundamental laws that govern and shape the development of cities.

These written materials were also valuable:

Bettencourt, L. M. A., et al. Growth, innovation, scaling, and the pace of life in cities. *Proceedings of the National Academy of Sciences of the United States of America* 104(17):7301–7306. www.pnas.org/content/104/17/7301.full.pdf+html.

Bloomberg, M. R., and Aggarwala, R. T. Think locally, act globally: How curbing global warming emissions can improve local public health. *American Journal of Preventive Medicine 35*(5):414–423, 2008. www.ajpmonline.org/article/S0749-3797(08)00705-8/fulltext.

Bragdon, D. The City of New York: Testimony for David Bragdon, New York City Council Oversight Hearing: Climate Change, December 16, 2011. www.nyc.gov/html/planyc2030/downloads/pdf/dbragdon_climate_testimony_111216.pdf.

Brody, J. E. Communities learn the good life can be a killer. Well [blog], NYTimes.com, January 30, 2012. http://well.blogs.nytimes.com/2012/01/30/communities-learn-the-good-life-can-be-a-killer.

Chappel, B. In America's 'most bikeable' cities, bike lanes rule. NPR.org, May 18, 2012. www.npr.org/2012/05/18/153029914/minneapolis-portland-ore-top-bikeable-cities-list.

DiMaggio, A. Suburban sprawl and the decline of social capital [op-ed]. Truthout.org, August 22, 2010. http://archive.truthout.org/suburban-sprawl-and-decline-social-capital62465.

Doig, W. Are freeways doomed? *Salon*, November 30, 2011. www.salon.com/2011/12/01/are_freeways_doomed.

Durning, A. Vancouver is Cascadia's greenest city, who is second [opinion]? TheTyee.com, January 7, 2010. http://thetyee.ca/Opinion/2010/01/07/CascadiasGreenestCity.

Edgington, D., and Goldberg, M. A. *Global and local forces at work in Vancouver: The fascinating case of the birth and rebirth of Coal Harbour.* U21Global Working Paper No. 017/2006, December 1, 2006. http://dx.doi.org/10.2139/ssrn.1606305.

Finch, K. Plan revamps NYC food systems. UrbanFarmOnline.com, February 25, 2011. www.urbanfarmonline.com/urban-farm-news/2011/02/25/plan-revamps-nyc-food-systems.aspx.

Flocks, S. 'I can't get there from here' [opinion]. AJC.com, January 9, 2012. www.ajc.com/opinion/i-cant-get-there-1296757.html.

Frumkin, H. Urban sprawl and public health. *Public Health Reports* 117(3):201–217, 2002. www.ncbi.nlm.nih.gov/pmc/articles/PMC1497432/pdf/12432132.pdf.

Glaeser, E. What a city needs. *New Republic*, September 4, 2009. www.tnr.com/article/books-and-arts/what-city-needs.

Goldberger, P. New York's High Line: Miracle above Manhattan. *National Geographic*, April 2011. http://ngm.nationalgeographic.com/2011/04/ny-high-line/goldberger-text.

Goldmark, A. NYC energy audit shows cleaner city, but not because of transportation. TransportationNation.org, September 19, 2011. http://transportationnation.org/2011/09/19/nyc-energy-audit-shows-cleaner-city-but-not-because-of-transportation.

Gray, C. As High Line park rises, a time capsule remains. *New York Times*, May 18, 2008. www.nytimes.com/2008/05/18/realestate/18scap.html.

Harcourt, M., et al. *City making in paradise: Nine decisions that saved Vancouver.* Vancouver, BC: Douglas and McIntyre, 2007.

Holland, M. The eight pillars of a sustainable community. HB Lanarc, n.d. www.hblanarc.ca/attachments/8pillars_matrix_HBL.pdf.

Izeman, M. NYC enacts new local food laws. Mark Izeman [blog], Switchboard.nrdc.org, August 17, 2011. http://switchboard.nrdc.org/blogs/mizeman/nyc_enacts_new_local_food_laws.html.

Jackson, R. L., and Sinclair, S. *Designing healthy communities.* San Francisco: Jossey-Bass, 2012.

Lehrer, J. A physicist solves the city. *New York Times Magazine*, December 17, 2010. www.nytimes.com/2010/12/19/magazine/19Urban_West-t.html.

Lerner, J. How urban planning can improve public health. *Pacific Standard*, April 28, 2010. www.psmag.com/health/how-urban-planning-can-improve-public-health-11408.

Levine, L. New report highlights vulnerability of NYC water infrastructure to climate change—and the city's efforts to prepare. Larry Levine [blog], Switchboard.nrdc.org, July 26, 2011. http://switchboard.nrdc.org/blogs/llevine/today_nrdc_released_link_to.html.

Levine, L., and Chou, B. New York and Pennsylvania: Among the best at planning for the inconvenient truths of climate change. Larry Levine [blog], Switchboard.nrdc.org, April 5, 2012. http://switchboard.nrdc.org/blogs/llevine/new_york_and_pennsylvania_amon.html.

Martin, G. New York's hanging gardens. *The Guardian*, November 7, 2009. www.guardian.co.uk/world/2009/nov/08/highline-new-york-garden-martin.

New York City Council. Speaker Christine C. Quinn delivers pioneering food policy speech [press release]. November 22, 2010. http://council.nyc.gov/html/pr/11_22_10_foodworks.shtml.

Owen, D. *Green metropolis: Why living smaller, living closer, and driving less are the keys to sustainability.* New York: Riverhead Books, 2009.

Padgett, T. Florida's deadly hit-and-run car culture. *Time*, November 29, 2009. www.time.com/time/nation/article/0,8599,1942986,00.html.

Reid, T. R. The superhighway to everywhere. *Washington Post*, June 28, 2006. www.washingtonpost.com/wp-dyn/content/article/2006/06/27/AR2006062701830.html.

Repko, M. Urban planner paving the way for bike-friendly Dallas. *Dallas Morning News*, September 12, 2010. www.dallasnews.com/news/state/headlines/20100912-urban-planner-paving-the-way-for-bike-friendly-dallas-.ece.

Rutkoff, A. NYC emerges as 'green city' leader. Metropolis [blog], WSJ.com, June 30, 2011. http://blogs.wsj.com/metropolis/2011/06/30/new-york-emerges-as-green-city-leader.

Sheraton, M. West 12th Street, by the numbers. *New York Times*, October 20, 2006. www.nytimes.com/2006/10/20/arts/design/20city.html.

Snyder, L. T. President Dwight Eisenhower and America's Interstate Highway System. *American History*, June 2006. www.historynet.com/president-dwight-eisenhower-and-americas-interstate-highway-system.htm.

Speck, J. Our ailing communities. MetropolisMag.com, October 11, 2006. www.metropolismag.com/story/20061011/our-ailing-communities.

US Department of Transportation Federal Highway Administration. Frequently asked questions: Interstate FAQ. n.d. www.fhwa.dot.gov/interstate/faq.htm.

Vancouver Sun. Some things worked. *Vancouver Sun*, September 6, 2007. www.canada.com/vancouversun/news/westcoastnews/story.html?id=ad56af4e-0f14-4717-9603-5fe5a0713e4c.

Vanderbilt, T. The crisis in American walking. *Slate*, April 10, 2012. www.slate.com/articles/life/walking/2012/04/why_don_t_americans_walk_more_the_crisis_of_pedestrianism_.html.

———. Learning to walk. *Slate*, April 13, 2012. www.slate.com/articles/life/walking/2012/04/walking_in_america_how_we_can_become_pedestrians_once_more_.html.

———. Sidewalk science. *Slate*, April 11, 2012. www.slate.com/articles/life/walking/2012/04/walking_in_america_what_scientists_know_about_how_pedestrians_really_behave_.html.

SELECTED BIBLIOGRAPHY

Carey, Nessa. *The Epigenetics Revolution: How Modern Biology Is Rewriting Our Understanding of Genetics, Disease, and Inheritance*. New York: Columbia University Press, 2012.

Cox, Stan. *Losing Our Cool: Uncomfortable Truths about Our Air-Conditioned World (and Finding New Ways to Get Through the Summer)*. New York: The New Press, 2010.

Cullen, Heidi. *The Weather of the Future: Heat Waves, Extreme Storms, and Other Scenes from a Climate-Changed Planet*. New York: HarperCollins Publishers, 2010.

Deichmann, Richard E., MD. *Code Blue: A Katrina Physician's Memoir*. Bloomington, IN: iUniverse Star, 2007.

Drexler, Madeline. *Secret Agents: The Menace of Emerging Infections*. New York: Penguin Group, 2002.

Egan, Timothy. *The Worst Hard Time: The Untold Story of Those Who Survived the Great American Dust Bowl*. New York: Houghton Mifflin Harcourt, 2006.

Freudenburg, William R., Robert Gramling, Shirley Laska, and Kai T. Erikson. *Catastrophe in the Making: The Engineering of Katrina and the Disasters of Tomorrow*. Washington, DC: Island Press, 2009.

Friedman, Thomas L. *Hot, Flat, and Crowded: Why We Need a Green Revolution—And How It Can Renew America*. New York: Farrar, Straus and Giroux, 2008.

Hansen, James. *Storms of My Grandchildren: The Truth about the Coming Climate Catastrophe and Our Last Chance to Save Humanity*. New York: Bloomsbury, 2009.

Harcourt, Mike, and Kent Cameron, with Sean Rossiter. *City Making in Paradise: Nine Decisions That Saved Vancouver*. Vancouver/Toronto: Douglas & McIntyre, 2007.

Hertsgaard, Mark. *Hot: Living through the Next Fifty Years on Earth*. New York: Houghton Mifflin Harcourt, 2011.

Horne, Jed. *Breach of Faith: Hurricane Katrina and the Near Death of a Great American City*. New York: Random House Trade Paperbacks, 2008.

Jackson, Richard J., MD, with Stacy Sinclair. *Designing Healthy Communities*. San Francisco: Jossey-Bass, 2012.

Klinenberg, Eric. *Heat Wave: A Social Autopsy of Disaster in Chicago*. Chicago: University of Chicago Press, 2002.

Krupp, Fred, and Miriam Horn. *Earth, the Sequel: The Race to Reinvent Energy and Stop Global Warming*. New York: W.W. Norton & Company, Inc., 2009.

Lynas, Mark. *Six Degrees: Our Future on a Hotter Planet*. Washington, DC: National Geographic Society, 2008.

McKenna, Maryn. *Beating Back the Devil: On the Front Lines with the Disease Detectives of the Epidemic Intelligence Service*. New York: Free Press, 2004.

Owen, David. *Green Metropolis: Why Living Smaller, Living Closer, and Driving Less Are the Keys to Sustainability*. New York: Penguin Group, 2009.

Pearce, Fred. *With Speed and Violence: Why Scientists Fear Tipping Points in Climate Change*. Boston: Beacon Press, 2007.

Pooley, Eric. *The Climate War: True Believers, Power Brokers, and the Fight to Save the Earth*. New York: Hyperion Books, 2010.

Stallings, Frank L., Jr. *Black Sunday: The Great Dust Storm of April 14, 1935*. Austin: Eakin Press, 2001.

Tidwell, Mike. *The Ravaging Tide: Strange Weather, Future Katrinas, and the Coming Death of America's Coastal Cities*. New York: Free Press, 2006.

Weis, Tony. *The Global Food Economy: The Battle for the Future of Farming*. New York: Zed Books, 2007.

INDEX